Psychometry and the Storage of Spiritual Energy in Crystals

Malcolm K. Smith, Ph.D.

Psychometry and the Storage of Spiritual Energy in Crystals

Copyright © 2013 by Malcolm K. Smith

All Rights Reserved.
Unauthorized duplication or distribution is strictly prohibited.

ISBN 10: 1933817755
ISBN 13: 978-1-933817-75-0

Cover design provided by Susan Veach based on an image by Klaus Monies.
Editing assistance provided by Kathryn Bartman.

Published by: Profits Publishing
http://profitspublishing.com

Canadian Address
1265 Charter Hill Drive
Coquitlam, BC, V3E 1P1
Phone: (604) 941-3041
Fax: (604) 944-7993

US Address
1300 Boblett Street
Unit A-218
Blaine, WA 98230
Phone: (866) 492-6623
Fax: (250) 493-6603

Acknowledgments

I must acknowledge the contributions of my human guides, Alannah Jantzen and Susan Harris, who have shown me pathways I never knew existed. Throughout the writing of this book Dawn Stewart has applied her editing skill and patience to help me say what I really mean.

Klaus Monies has been a great help as a source of illustration suggestions, and I am glad we were able to include six of his diagrams in the final book.

Thank you my dear human friends.

Contents

Introduction 1
Chapter 1: Storage of Spiritual Energy in Crystals 5
Chapter 2: Storing Energy in Semi-Crystalline Materials ... 21
Chapter 3: Ghost and Spirit Energies 33
Chapter 4: Storing Energy in Liquid Crystals 45
Chapter 5: Cosmic Lattice Psychometry 55
Chapter 6: Homeopathy and Crop Formations 65
Chapter 7: Healing with the Electromagnetic Song of Atoms and Molecules 77
Epilogue 89
Appendices 1-16 93
Index 241

Introduction

With Judy's words, "Just trust!" ringing in my ears, I took the metal pendant into the deserted lunch room, sat at a table and started to write. As I had picked up the spiral metal pendant from the collection of items offered for testing, I had "heard" two words, "artistic" and "sensitive," so now I wrote those words down as my first impressions.

I held the pendant in my left hand as I wrote with my right. The impressions came fast – like a picture-in-picture TV screen in my mind. In the inner picture I saw:

- An impression of the family of the owner of the pendant (who was unknown to me at this time). I saw her siblings and mother and an unusual relationship of these people to the owner. A feeling of "not completely belonging."
- Through the eyes of the owner, I saw a meadow in springtime. I felt this time was especially important to her, but the joy was dampened by feelings of loss.
- Also through her eyes, I saw a seaside place with a pier jutting out from a promenade on which she walked. I felt this place was England, and the owner was there looking for her family.

I was in a psychometry class, learning to use what my dictionary calls the "supposed" faculty of humans to read information about people from objects that have been in their possession. I returned from my quiet corner in the lunchroom to the rest of the class. When my turn came I gave my reading as recorded above. As I finished Alison revealed herself as the owner of the pendant with the words, "You are 100% accurate." She explained:

- She had been born in England, but was given up for adoption by her birth mother, who married and had five more children.

- Her birthday was in April, but her mother was always sad at that time for her loss.
- Alison had been to a seaside town in England recently searching for, and meeting, her birth family.

From our knowledge of Alison, we all recognized the accuracy of my first impression words: "artistic" and "sensitive."

This experience amazed me. I found it surprising that so much accurate information could be available to me, an absolute beginner in psychometry. I had been able to read the metal of that pendant like looking at pictures in a book. I resolved to find out where and how those pictures were recorded in the metal. This book tells the story and results of my research.

How I get my information:

I am a chemist by training and a spiritual researcher by inclination. This book is based on channeled information combined with concepts derived from experimental science. The channeled part comes from two sources:

- A group of Angels called "Crystal Light" give me words to say which I record and later transcribe into writing.
- Spirit Guides give me words via a pendulum and an alphanumeric chart. The Guides arrange the channeling sessions, search out additional information and explain obscure points to me as well as guiding me on my life path.

On occasions, the two techniques are combined – for example, in checking by pendulum words that were not clear in a recorded channeling.

Some information is added from other channelers and authors. In those cases, the source is acknowledged by references.

For details of this procedure, I refer you to my first book: *Spiritual Chemistry – The Interaction of Spiritual Energy with the Physical World*.

Chapter 1: Storage of Spiritual Energy in Crystals

In this chapter:

- ❖ *The Structure of Physical Materials*
- ❖ *Crystals in History*
- ❖ *Special Powers of Crystals*
- ❖ *Minerals*
- ❖ *Metals*
- ❖ *Crystal Structure*
- ❖ *Crystal Versus Glass*
- ❖ *Unit Cells*
- ❖ *The Seven Crystal Systems*
- ❖ *Storing Energy in Crystals*
- ❖ *The Musical Notation Analogy*
- ❖ *Golden Angel, Oversoul or Higher Self*
- ❖ *Reading the Energy*
- ❖ *Crystals in Electronics*
- ❖ *Crystal Skulls*

The Structure of Physical Materials

All physical materials are made of atoms. In some solids, for example, metals, the atoms exist singly, but are linked to many others in crystals. In other solids, for example, ice, groups of atoms are bonded together to form molecules, which are linked to many other molecules in a crystal. (In ice one oxygen and two hydrogen atoms are bonded together to form each water molecule – hence the formula H_2O.)

Crystals are the most perfect arrangement of atoms or molecules. I will describe their structure and energy-holding capacity in this first chapter. However, there are many other materials, such as wood or bone, which are only partly crystalline. You can learn about semi-crystalline materials – including liquid crystals – in later chapters.

Crystals in History

For millennia, crystals have been regarded by humans as magical because of their translucency and regular shape, usually with flat faces. Because of their apparently supernatural shape, crystals appear to have been made artificially, and in the distant past fantastic stories developed about their creation. For example, Pliny the Elder (Ref. 1-1) – a Roman writing in the first 100 years after the birth of Jesus – described stones that originated in the stomachs of animals or grew in the light of the moon. He asserted that all these stones possessed magical powers, such as the ability to quiet winds or increase intelligence of the owner.

Pliny recognized several types of stone, including one we now know as quartz, which he called crystal after *krystallos*, the Greek word for ice. Like many people of his time, he believed this stone occurred only in mountainous areas, like the Alps, and was water that was frozen so hard it would no longer melt.

In the light of knowledge available at that time, this seems to me to be a smart guess by Pliny. He wasn't to know, as we do now after several hundred years of physics and chemistry discoveries, that the regular shape of crystals was the result of them being formed by billions of atoms or molecules regularly packed in rows and sheets, which are stacked layer upon layer. We will return to the structure of crystals later in this chapter and see how it gives them the ability to store energy and information within their layers.

Special Powers of Crystals

In Pliny's time – 2000 years ago – magical powers, such as quieting storms, were accorded to crystalline stones, which I will refer to mostly by the general term, minerals. Starting with these early beliefs, trial and error methods have gradually built a body of knowledge collected from experiments without reference to any theories. In modern times, that body of empirical knowledge has been collected into books, which act as encyclopedia of the beneficial properties of minerals for humans and animals (Ref. 1-2).

For example from the knowledge in (Ref. 1-2) I made a black tourmaline necklace for a relation of mine when she was diagnosed with breast cancer. I believe that mineral contributed significantly to her successful recovery. Some scientifically trained people – who concede there may be something in this approach that rises above superstition – usually attribute this kind of result to a placebo effect. They say that the patient's belief that wearing the mineral improves their health is responsible for the healing. I agree that the placebo effect may be partially responsible for the healing. But we shall see later there is a physical basis for the beneficial effects of minerals which is related to homeopathy. I know many medical practitioners still do not believe homeopathy can heal humans, but some agree there is an effect worth studying (Ref. 1-3). Later in this book, the Crystal Light Angels and I will provide an explanation for the success of homeopathy and crystalline mineral cures.

Minerals

Minerals can be divided into two groups. There are obviously crystalline ones in which flat crystal faces can be seen, for example, quartz. Then there are apparently amorphous minerals which look like a collection of sand particles stuck together with something. But I assure you the amorphous minerals are nearly all aggregates of small crystals – if you cut a sample of agate, polish the cut surface and look at it with a magnifier, you will see crystal grains. There are some exceptions, for example, obsidian volcanic glass. I will show you later the difference between glasses and crystals in a special section. So, in conclusion to this section, we can state that practically all minerals are made of crystals. A later section starts a fascinating story of how crystalline materials can store energy, but first we must look at the structure of crystals at the microscopic level.

Metals

Although metals do not look crystalline, they usually consist of many small crystalline grains – like agate in the previous section. As molten metal cools and solidifies, the atoms quickly arrange themselves into crystals. The crystals start growing – scientists say nucleate – at many points in the cooling metal. Each crystal keeps growing until it meets another crystal, and then the two crystals form a common grain boundary. Eventually the whole metal object consists of a mass of interlocking crystals, something like a three-dimensional jigsaw puzzle.

Crystal Structure

Crystals are characterized as having flat faces set at definite angles to each other. The perfectly flat faces are the result of the regular arrangement of atoms that make up the crystal. All the atoms inside the crystal are arranged in parallel rows like soldiers on parade. The parallel rows form sheets of atoms – which are very big in relation to the size of the atoms – and the sheets form layer upon layer. See Figure 1-1. So, if we could look inside the crystal with a very powerful microscope, we would see a three-dimensional grid of uniformly-spaced atoms which is very big in relation to the size of the atoms, so big that we are aware of its existence in our world.

Not only single atoms form crystals. There are also crystals composed of molecules which are groups of atoms bonded together. Probably the most familiar example of that is ice in which water molecules – consisting of two hydrogen atoms bonded to one oxygen atom (H_2O) – line up to join in a growing crystal when their temperature dips below freezing point.

While I am mentioning chemistry, you should know that some crystals are formed from ions in rows and sheets. An ion is an atom that has gained an electron – so it has a negative charge – or has lost an electron – so it has a positive charge. For example, salt crystals are made up of positively charged sodium ions and negatively charged chlorine ions.

As a crystal grows, new atoms (or molecules or ions) line up in the solution or melt to take a place in the growing crystal grid, known as a lattice, because this represents a state of low energy for all the atoms. As they approach their neighbours in the grid, the atoms experience inter-atomic forces of repulsion from energy in the atoms' structure. This stops them at a small distance from each other and results in the very uniform spacing between all the atoms in the crystal.

Crystal Versus Glass

This seems to be a good point to deal briefly with glasses. When a molten material is cooled too fast for the atoms or molecules to take their place in the growing crystal grid, the result is non-crystalline glass in which the atoms are arranged randomly. It is confusing that we have in the modern world expensive lead-glass vessels and ornaments that are referred to as crystal. But this use of the incorrect name, "crystal," is still with us because many of these objects used to be made from clear quartz, e.g. fortune-tellers' crystal balls.

You may have heard of obsidian, which is a naturally occurring volcanic glass formed when molten lava cools too rapidly to allow the molecules to form crystals.

Unit Cells

In crystals of common salt – chemical name sodium chloride – the ions of sodium and chlorine repel each other with the same force in all three directions. The result is they lie at the corners of a series of imaginary cube-like structures as shown in Figure 1-2. In a real salt crystal, millions of cubes like it – with ions at each corner – are stacked together to form a salt crystal, which ideally is in the form of a cube (more on deviations from the ideal shape later). So we can regard this basic cube, with 8 ions at the corners, as the building block of the salt crystal. For this reason, this basic cube is called the unit cell of the salt crystal.

If you look at salt crystals under a magnifying glass, you will see that very few are exact cubes. For example, if the concentrations of ions around the growing crystals were unequal – perhaps they were more concentrated at the top of the growing crystal and less concentrated at the sides – that situation would lead to elongated crystals which would appear like needles. But because of the cubic unit cell the crystals would still have faces at right angles (90°) to each other.

The Seven Crystal Systems

In salt crystals, the forces between simple round sodium and chlorine ions are the same in all directions, and this gives rise to a unit cell with equal sides, as shown in Figure 1-2. However, most molecules are not simply round, they have many shapes. Molecules have different shapes depending on the chemical compound they are made from. For example, dye molecules are often shaped like flat discs, and so the forces between the flat faces are different to the forces between the edges of the molecule. This may lead to unit cells that are elongated, and in some cases with sides that are at angles other then 90° to each other.

Crystallographers have been studying the possible shapes of naturally occurring crystal unit cells for several hundred years.

They have isolated seven basic forms with names like triclinic and tetragonal. If you are interested in following up on some fairly complicated three-dimensional geometry, then I refer you to books like *Mineralogy* (Ref. 1-1). But we have covered crystal structure sufficiently well to be able to see that unit cells can act as little cages or buckets for storing energy. I will discuss that concept in the next section.

Storing Energy in Crystals

As we discussed earlier, in crystals, the atoms, or molecules, are arranged in a regimented way – like soldiers on parade, except this parade is in three dimensions. All the atoms stand at similar distances from each other and form a very regular three-dimensional matrix. In this matrix, the atoms vibrate a little about their average position but generally they don't move much, so that the crystal is manifested as a solid object. (The atoms don't move significantly until their energy increases with temperature to a point where the crystal melts.) In this stable matrix, each unit cell forms a sort of frame which allows energy to come in and be held vibrating in the frame.

The Crystal Light Angels say: "An analogy which comes to our mind that would make sense to you on Earth is that these crystal atom grids are like picture frames. The energy of the atoms of the crystal form the frame itself – this is a three-dimensional frame, not two-dimensional as you would have on a wall in your world – and this three-dimensional frame is quite stable and strong and is able to hold a lot of information as energy. That information energy comes in from outside sources, say from some event that occurs around the crystal, and the energy is then held like the picture in the picture frame. It is held between the energies of the atoms around it forming a very stable cell." (Appendix 2)

The Music Notation Analogy

Another way of looking at the storage of information energy

in a crystal is to consider the crystal's unit cells as measures in printed music notation. The atoms are equivalent to the staff and bar lines creating a frame in which energy can be stored. In this analogy, the stored energy is represented by the music notes that are written on the lines of the staff. Each measure of the music – or bar – represents one unit cell of the crystal. Then, in each measure of the music we have a two-dimensional version of a crystal's three-dimensional unit cell. Into that two-dimensional musical space can be put concepts, in the form of notes and rests, which become the information telling musicians how to play that particular section of the music and recreate its energy.

The Angels say that this is a very good analogy because it has the feeling of energy coming from the music. It may appeal to many people, particularly the non-scientific ones, who may not be sure of the structure of crystals, but can see how the bar lines divide the music up into measures equivalent to unit cells in a crystal.

Humans are able to read out the information energy in the unit cells of a crystal. That is the basis of psychometry. But of course the big question is, "How do we do that?" First, let me explain about each human's Golden Angel.

Golden Angel, Oversoul or Higher Self

After reading Kryon (Ref. 1-4), talking with my Guides and being aware of my own Golden Angel, I have come to the conclusion that the following statement from my first book (Ref. 1-5) is correct:

All humans have an Angel and at least two guides available to help them.

I am now going to tell you about the relationship between the Angel and each human. First let's talk about the Angel and other names by which they are known:

- Golden Angel – with the same face as the human – Kryon's description (Ref. 1-4)
- Oversoul – a name from Seth/Jane Roberts (Refs. 1-6, 1-7)
- Higher self – with connections to psychology

The total energy of our Golden Angel, or Oversoul, is far too great to be able to exist in a single human body. For that reason the Angel sends several parts of itself, called aspects (Ref. 1-6), into the Earth plane as humans. I am told that up to 7 aspects can be sent by each Angel. In the case of my inter-dimensional "family," my Golden Angel Mikael has sent three aspects into the Earth plane.

In the following section I shall use only the name Golden Angel with the understanding that the other two names can be substituted if you prefer them.

Reading the Energy

Humans are able to read out the energy that is stored in the crystals without removing the energy from the crystals. It's like reading a book in a library – you don't have to remove the book from the library. You can take the book off the shelf, read the information that is stored in the book, and put it back on the shelf. Reading information energy stored in crystals is the technique known as psychometry, which my dictionary describes as the "supposed" faculty of humans to read information about people from objects that have been in their possession, e.g. jewelry and clothing.

The question is how do we read out that information stored in the crystal? I will let the Crystal Light Angels answer that in their own words (From Appendix 8):

> "The first point we want to make is that all of this reading is done with the help of your Golden Angel. In all

cases the Golden Angel – Mikael in your case Malcolm – goes into the crystalline object, e.g. a metal pendant, and brings out the information that is required. It seems like the person doing the psychometric reading is actually reading the energy out of the crystal but in fact that process is greatly facilitated by the Golden Angel of each person doing the reading."

"In your case Mikael always assists you in reading out the energy. He goes into the crystal and, as it were, decodes the symbols that are within each unit cell of the crystal and brings them back to you – telepathically of course, there's no actual traveling to and from the crystal. His intent goes into the crystal, finds the necessary information and feeds it back to you. That is why when you do psychometry and you have the words about the owner of the object, it seems as though those words come from some other person or source. That is in fact true because the words come from your Golden Angel Mikael. He is facilitating the reading process and passing the information to your spirit who relays it through your body's DNA by magnetic impulses to your brain where it becomes conscious. You were able to read out that information just as if it were words in a book because Mikael was translating the information from the crystals into your consciousness."

To reach that information, the Golden Angel makes use of the cosmic lattice – an energy and information construct throughout the universe – which will be described in Chapter 5. The cosmic lattice runs through the crystals, and can be considered the conveyor that brings the information in the first place into the crystal where it can be stored. The interaction of the cosmic lattice and each human's multidimensional DNA will also be described later in this book.

Continuing the Angels' advice from Appendix 8:

"For anyone who is wishing to learn how to do psychometry it is preferable for them to obtain the help of their Golden Angel. Now this is where it divides into two streams:

- In the first case the person doing the psychometric reading – we will call them the reader – is aware of their Golden Angel, as you are Malcolm, and they acknowledge the fact that the Golden Angel is helping them.
- In the other case the reader does not know about the Golden Angel. All that person knows is that it <u>seems</u> the information is given to them when they focus their intent on the object which they are reading.

"That's how all psychometry works, either the reader is not aware of the Golden Angel's help or the reader is conscious of the Golden Angel's presence and help. In our opinion, in the case of the conscious awareness of the Golden Angel, the skill of psychometry is more reliable, because the intent of the Golden Angel is recognized and that strengthens the link between the Angel and the reader."

Crystals in Electronics

You may have heard of crystals being used in electronic information technology. For example, quartz crystals are used in a time-keeping circuit in watches and clocks, and silicon and germanium crystals are employed in semiconductor devices – such as transistors – which are used widely in computers and many other electronic products. Although these devices are the mainstay of information technology, the crystals in them do not hold information in their unit cells as I have described so far in this chapter. The watch time-keeping circuits make use of quartz crystals' piezoelectric effect, which causes the crystal to vibrate when fed with a small electric current (Ref. 1-8). In the case of semiconductor devices, electrons – from the crystal atoms – are manipulated through junctions of two dissimilar crystals to give a positive or negative charge either side of the junction. The crystal junctions – transistors – can be used to manipulate information and even store it on a temporary basis, but these devices cannot store information indefinitely as the unit cells of crystals can.

Crystal Skulls

This discussion of information storage in crystals would not be complete without a brief review of some ancient crystal devices which are believed to be capable of storing large amounts of information for millennia. These are the crystal skulls found in North, Central and South America and which range in size from full human skull to miniature versions. It is generally recognized that there are 13 skulls (Ref. 1-9) which are claimed to be from ancient sources, although there is no way of authenticating their age.

The Mitchell-Hedges skull was found in a Mayan ruin in 1924 by Anna, the 17-year-old daughter of Frederick Mitchell-Hedges. Scientific tests confirm that the skull and detachable jaw bone have been made from a single block of quartz crystal. Since quartz is only slightly less hard than diamond, archeological experts assume that in the absence of sophisticated grinding technology it "must" have taken the craftsmen 300 man-years of toil to fashion the skull by grinding the quartz block with particle-size graded sands. However even this labour-intensive method of crafting the skull is doubtful, because microscopic examination of the skull's surfaces cannot detect any scratch marks. It appears that the only way to explain the perfection of the skull is to consider it the product of superior alien technology. Mitchell-Hedges himself appeared to be thinking along the same lines, because he suggested the skull was Atlantean in origin, particularly when he saw how the Mayan people treated it as a very sacred object holding ancient wisdom.

These suggestions about the fabrication of the skulls have been confirmed in Appendix 9 by the Crystal Light Angels as follows:

> "We could say that the crystalline grid – a network of magnetic lines around the Earth – is humanity's special corner of the cosmic lattice. This is the part that is connected with the crystal skulls. We know, Malcolm, you developed your interest in them when you read the book about the skulls (Ref. 1-9). It was apparent to you that there was a connection with the crystals you were talking about in Chapter 1 of our new book and that the energy that is stored in them is paralleled by the energy stored in the crystal skulls. You know now, as a result of reading Kryon (Ref. 1-10) that the crystalline grid is anchored to the physical body of the Earth by the crystal skulls. The grid was put in place at the time of Lemuria and later modified in the time of Atlantis. Those crystal skulls were made by matter creation in the shape of the human skull. Just as Sai Baba (Ref. 1-11) can materialize physical objects in the modern world, so the Lemurians and some of the Atlanteans could materialize objects and that is how those crystal skulls came into being. The purpose of them is to anchor the crystalline grid to the surface of the Earth and act as a 'road map' for humanity's development (Ref. 1-9). Also they record past events in the very long life of humanity on the Earth which goes back 100,000 years (Ref. 1-12), to the time when people from the Pleiades came and interbred with humans and changed the human multidimensional DNA to include 'the search for God.'"

In case you're wondering if the skulls should be put back in their original hiding place to continue acting as anchors, Kryon tells us (Ref. 1-10) that when one is found and pulled out of the ground, it ceases to work as an anchor. It is simply a historical object with very special spiritual energy. So if you find one, let it remain where it is.

In the 1990s, when the book about the skulls (Ref. 1-9) was written, the Mitchell-Hedges skull was kept at the home of Anna Mitchell-Hedges in Kitchener, Ontario, Canada. The authors of the book were present during a session in which Carole Wilson – a Canadian psychic and police psychometry expert – accessed the information in the skull. In a trance state Carole revealed a very large amount of information about the skull and the reason for its existence, part of which confirms the time and method of creation of the skull as described by the Crystal Light Angels.

In conclusion I want to underline something the Crystal Light Angels said: The energy that can be stored in crystals is immense. The crystal skulls not only anchor the crystalline grid – which is described in detail in Appendix 9 – but they also record past events of humanity's 100,000 year life on the Earth and hold a road map for its future development.

Author's Note: For Angel advice on removal of unwanted energy from any crystals you may have, I refer you to Appendix 2. – MKS

References for Chapter 1

Ref. 1-1 *Mineralogy* by L.G. Berry, B. Mason & R.V. Dietrich

Ref. 1-2 *Healing Crystals and Gemstones* by Flora Peschek-Böhmer & Gisela Schreiber

Ref. 1-3 *The Field* by Lynne McTaggart

Ref. 1-4 *Kryon Book 3 – Partnering with God* by Lee Carroll

Ref. 1-5 *Spiritual Chemistry* by Malcolm K. Smith

Ref. 1-6 *Adventures in Consciousness* by Jane Roberts

Ref. 1-7 *The Education of Oversoul 7* by Jane Roberts

Ref. 1-8 *Physics Today* Published by World Book, Inc.

Ref. 1-9 *The Mystery of the Crystal Skulls* by Chris Morton & Ceri Louise Thomas

Ref. 1-10 *Kryon Book 7 – Letters from Home* by Lee Carroll

Ref. 1-11 *Miracles – A Parascientific Inquiry into Wondrous Phenomena* by D. Scott Rogo

Ref. 1-12 *Kryon Book 12 – The Twelve Layers of DNA* by Lee Carroll

Chapter 2: Storing Energy in Semi-Crystalline Materials

In this chapter:

- *Semi-Crystalline Materials*
- *Information Energy Stored in Semi-Crystalline Materials*
- *Crystals in Bone*
- *Information Energy Stored in Bones*
- *Energy Stored in the Bones of the Saints*
- *Pilgrimages to Share the Energy of the Saints*
- *The Story of Santiago de Compostela*
- *My Visit to Santiago de Compostela*

Semi-Crystalline Materials

In Chapter 1, I have been talking about materials that are completely crystalline. They consist of either large single crystals – like quartz – or aggregates of crystal grains – like agate. Although they do not usually appear crystalline, most metals consist of crystal grains too. The substances that form these relatively large crystals consist of single atoms, small or medium-sized molecules. The latter are made up of about 100 atoms, but in spite of their size they can move about quickly when melted or in a solution. It is the mobility of the single atoms and the small to medium-size molecules that allows them to form large crystals. This is because at the point in the growing crystal where new atoms or molecules are adding themselves into the regimented rows of atoms or molecules, they have little time to orient themselves to fit into the growing crystal. It's like a game of musical chairs when the music has just stopped and the atoms have to rush to find a forming unit cell that they can "sit" in.

Now I am going to talk about giant polymer molecules – consisting of thousands of atoms – which make up such familiar materials as wood, paper, textiles and plastics. Usually these molecules are in the form of chains that are so long and slow-moving that only part of each molecule can fit into a crystal. In fact, several different sections of the same molecule may take part in contributing to several crystalline areas. These materials are referred to as semi-crystalline, and typically have structures similar to that shown in Figure 2-1. Notice how some of the molecules – shown as linked blocks – "snake" around and take part in several different crystalline areas. This is what scientists refer to as the "fringed micelle" model of polymer structure (Ref. 2-1). I like that name because it suggests little "pellets" of crystalline polymer with polymer chains protruding out of the sides of the pellets like "fringes."

Information Energy Stored in Semi-Crystalline Materials

What we learned in Chapter 1 about information and energy – which are essentially the same thing – being stored in materials made entirely of crystals also applies to semi-crystalline materials. For example, in the case of wood, plants, cotton, linen and rayon the crystals are composed of parallel parts of long cellulose molecules. Other polymeric materials used as plastic articles and synthetic fibres can also exhibit crystalline areas where the polymer molecules lie side by side. We can see that there are enough crystals in these materials for energy to be held within them just as in completely crystalline materials. This is where police psychics go – or rather their Golden Angel goes – when they want to make a psychometric reading on a missing person's clothing.

Crystals in Bone

Crystals in wood and plastics occur almost "accidentally" where molecules happen to lie side by side with other molecules. However, the crystalline areas of bone are not accidental; they are strategically placed to develop the strength and rigidity of bone that all animals, including humans, depend on for movement.

There are two types of human bone tissue: compact outer bone which is dense like ivory, and an inner network which encloses cell-filled spaces in a honeycomb structure like coral. This combination of structures gives strength, rigidity and also lightness to all the major bones in our bodies (Ref. 2-2).

Most of our bones began as rods of cartilage that acted as a temporary skeleton while we were still in the womb. After birth, the cartilage hardened into bone as mineral materials were laid down in the growing tissue. The mineral material consists of minute crystals of calcium compounds arranged in layers in the bone. The minerals are laid down by special cells – called osteoblasts – that have invaded the cartilage. As they work, these cells "wall themselves in" and become immobile bone cells, communicating with their neighbours only through tiny pores.

If you are interested in a little more chemistry, the calcium compounds that are laid down as crystals in the ossification – hardening – process, consist of a group of compounds called apatite. They are complex calcium phosphate compounds with the general formula $Ca_5(PO_4)_3X$ where X in bone is usually a hydroxyl – OH – ion (Ref. 2-3) but may also be a fluorine atom, for example in teeth.

The Angels continue the explanation: "Of course, as you know from your dairy produce advertisements, calcium is the main element that occurs in bones. Crystals of compounds of this material reside in the bones and form their main structure. Of all the body structures, they are the strongest and most resistant

23

to crushing and tearing. The skeleton serves a double purpose of maintaining the body, which you use as a vehicle, in its full position so that it can move around and function properly. In bones there are several different compounds, but the one we want to focus on today is apatite which forms crystals right throughout the human skeleton." (Appendix 3)

While we are talking about crystals in bone, I must mention crystals in the surface enamel of our teeth. Its composition is 96 percent mineral, notably small crystals of calcium phosphate and calcium fluoride contained in a fibrous protein network. Although prone to decay during life, after death teeth resist decay longer than any other body tissue, including bone (Ref. 2-2).

Information Energy Stored in Bones

We saw above that bones are semi-crystalline materials containing many apatite crystals. The Angels continue the explanation: "In addition to conferring strength on bone, apatite crystals have a second purpose of storing energy. This is the premium storage place for energy which is created and received by the human spirit during a lifetime. Most of the energy comes into the body via the DNA. The human spirit itself, which resides in the body contributes much of the energy that goes into the body. Some of the energy goes to create the body, maintain its structure and grow new cells – the turnover of cells is necessary for the good health of the body. There is a certain component of the energy that comes in, which is a recording of the experience of the body, and that recording goes into the crystals, which occur in the bones of the skeleton. That is the principal storage place for the energy of the experience of the human during a lifetime."

> "Remember our analogy that each measure of music can represent one unit cell of a crystal? Forming that unit cell are atoms and the energy of the atoms creates the cage in which other energy can be placed, can be read

out or can be removed ... So in bones, human bones, there are many, many, many of these musical measures that can store a lot of musical notes, representing the information energy that has come to the human by the way we just described, through the DNA and the human spirit residing in the body, creating the memories and the knowledge that are held in the human body. While a lot of this is stored in the human's physical brain, there's still a lot more stored in the bones of the human, within the crystals inside the bones. The bones can be regarded as the repository of much of the experience of the person during their lifetime." (Appendix 3)

Energy Stored in the Bones of the Saints

The Angels continue the story from Appendix 3:

"The other aspect that we want to talk about is that while all humans store energies or knowledge in their bones, in the majority of cases the information that is stored in the bones is not very great. Whereas in the case of the saints their lives were so blessed that much more information energy was stored in their bones. We refer to the example of Saint Teresa of Avila, who led such a blessed life and who went into religious ecstasy on a regular basis – so much so that she levitated and was known as the flying nun. Her case is typical of many of the saints who led such blessed lives that the energy they absorbed into their bones was much more than that of normal humans, and as a result that energy stayed in the bones after the passing of the spirit of the saint back to our realms. So the bodies of the saints did not decompose for many years, hundreds of years in the case of Saint Theresa. The bodies were preserved by the sheer energy that was stored in the bones. That energy allowed the body, minus the spirit, to continue to exist in a relatively stable physical form with no deterioration taking place for a long time."

"Eventually the body did deteriorate because the energy weakened or was removed. – We are told that the energy was removed from the body some great length of time after death – The energy was required for other lifetimes and was recalled by the spirit that put the energy into the body in the first place. That is why the bodies eventually decomposed, as the energy was taken back into other realms. But as a demonstration of the energy that was in the bones, the bodies did not decompose for several hundred years. Church people observed these things, marveled at them and regarded them as a miracle. It is a miracle, but you are beginning to learn the mechanisms behind the miracles and how those bodies can continue to exist without decomposition. That process is continuing with modern saints who have died – Mother Teresa is an example of that, her body will not decompose for some time. (Strange coincidence that her name is the same as the saint in Avila – is it not?)"

Summarizing this concept which will be new to humans:

"There is a certain component of the energy that comes in which is like a recording of the experience of the body, and that recording goes into the crystals, which occur in the bones of the skeleton. That is the principal storage place for the energy of the experience of the human during a lifetime. That is why the bones of saints are so revered more than any other part, because the bones contain the most vibrations, the most energy that has come from the lives of the saints. Some people in present times and many people in past years were able to detect the presence of energy stored in the saints' bones. That is why the collection of saints' relics – they were called relics but basically it was the bones – was such an important part of religious practices, particularly in medieval times." (Appendix 3)

Pilgrimages to Share the Energy of the Saints

"The pilgrims that came to these places – churches where the relics were kept – wanted to be able to pass close by the bones, because then they could experience some of the energy that was stored in the bones. They could 'read out' some of the energy, and thereby receive a blessing from the knowledge that was stored in the bones. This was the main value of collecting bones and relics of the saints. This was so the energy that the saints had accumulated could be passed on to their – we were going to call them admirers, but let's stick to the word pilgrims. A definition of pilgrims could be: people that wanted to share in the energy of the saints."

"As you know, many pilgrims went to places like Santiago de Compostela (Northwestern Spain) just to stand near the bones – they were prevented from touching them by the church authorities – but the bones were always placed in a location where people could pass very close to them. Malcolm, you have done that yourself, you have passed very close to the bones of Saint James in the cathedral of Santiago de Compostela. You know there was a special place under the altar where people could pass within a few centimetres of the bones and you have done that yourself. You didn't realize as you did it in this lifetime, but you were making a connection with your past lifetime when you were involved in securing those bones and bringing them to Santiago." (MKS – I remember that this act of passing close to the bones felt special – Please see a later section of this chapter **"My Visit to Santiago de Compostela"** for my experience in the cathedral.)

"The pilgrims were also looked upon as vacationers, because to go on a pilgrimage was a great adventure and involved much travel and seeing other parts of the world.

So it was a very popular thing in medieval times for people to do. As you know that practice still continues to this present day, people walk across Northern Spain and visit Santiago and share in that pilgrimage. However, for those modern pilgrims, the journey is a more important part of the ceremonies than the connection with the bones, because the value of the bones is not realized. The energy within them is not experienced to such a great degree, because the faith of the people in modern times is different to the faith that was held by the medieval pilgrims. We think medieval pilgrims got more from their experience of being near the bones than from the journey, whereas in modern times the journey, and the effort in making the journey, is the driving force for the pilgrimages that happen now." (Appendix 3)

The Story of Santiago de Compostela

Santiago de Compostela is in the Northwestern tip of the Iberian peninsula. The name in old local Spanish/Latin translates as "Saint James of the field of the star." I will let the Angels tell the story (Ref. Appendix 3), as it involved me in a long past lifetime:

> "Malcolm you had a wish to go back to Santiago de Compostela in Northwestern Spain – as you know the bones of Saint James (Santiago in Spanish) are stored in the cathedral there. That is why that place is so revered and why you are attracted to it, because you were involved in the collection of the bones and putting them in that cathedral. As you know the legend has it that there were shepherds in a field and a star came down into the field, hence the name Compostela – field of the star. Of course in modern times, you realize that star was a UFO, or an alien space ship if you prefer a more definitive term. The aliens were cooperating with Angels like us, the Angels

that are speaking through you now, and we showed those shepherds where Saint James's bones had been hidden."

"The reason for the hiding of the bones was that they were on their way from the Holy Land to Europe, and pirates attacked the ship that was carrying them across the Mediterranean. The crew of the ship realized that the pirates were coming, and so a small detachment left with the bones in a fast small boat that made it to shore before the pirates got to the ship. That small detachment of the crew hid the bones in a safe place. The Angels and the aliens knew of the hiding place, and when all was calm and worldly threats were gone, the Angels arranged this meeting and reparation to the shepherds in the field of the star."

"The craft was piloted by 'Nordic' aliens and was just like the ones that you and Marjorie (Ref. 2-4) traveled in. That alien space craft came down in the field and the Angels told the shepherds about the location of the bones. The shepherds went to the hiding place, recovered the bones and took them to the place that is now called Santiago de Compostela. Of course, the shepherds were guided by the Angels to do this. In those days it was regarded as a miraculous event – it would still be regarded as a miraculous event in modern times. In some ways it resembled the miracle of Fatima."

"You, Malcolm, were steeped in those occasions, those meetings with aliens. The reason you are so interested is that you were one of the shepherds that was involved in the recovery of Saint James's bones. One of your 'missions' in different lifetimes is to recover relics and to care for them. Now in this lifetime your scientific side is coming to the fore and you are explaining why they are so important and the understanding behind these events."

My Visit to Santiago de Compostela

Although I didn't realize it at the time, I experienced the love energy from the bones of Saint James in 1996. About two years before that date I was looking at books on Spain in a library when I was given – by the Library Angel – a book on Santiago de Compostela. I was immediately filled with a strong desire to visit this place that I had never heard of before, but which suddenly seemed so familiar and meaningful to me.

In 1996, my wife and I went on a bus tour of Portugal and Northern Spain. After driving north through Portugal, we arrived in Santiago de Compostela in the Northwestern corner of the Iberian peninsula. On our second day there, we went on a guided tour of the cathedral. Our guide told us the story of the supernatural recovery in AD 813 of the bones of the apostle Saint James who was martyred in Jerusalem about AD 44. Of course I did not know then of my previous lifetime involvement in the recovery, but the guide's mention of the star being a UFO caught my attention immediately, and a curious feeling of familiarity began stirring within me.

The highlight of the tour was an opportunity to pass close by the bones of Saint James or Santiago as we were referring to him by then. We were shown a narrow passage under the main altar and invited to line up with the present-day pilgrims to pass through the passage. This I did. Halfway along the passage was a small round window of very thick glass. When my turn came to look through the window I could see close to the window baskets of metal, which I guessed contained the bones, although I could not make them out clearly. But what impressed me was how I felt for those few seconds in front of the window – a great peace enveloped me and stayed with me as I continued down the narrow passage back into the magnificence of the great cathedral.

My wife and I had been assigned a room on the top floor of the hotel which had a skylight. Although it seemed a little unusual we assumed it was typical of Spanish hotels which were frequently in old buildings. During the night I went to the bathroom, and as I came back to bed I saw perfectly framed in the skylight the constellation of Orion. This struck me as very significant for two reasons: (a) I was considering Orion as the name for my consulting company that I was in process of setting up. (b) I had read in Graham Hancock's book (Ref. 2-5) about the significance of Orion for positioning the great pyramid. In addition to the significance of (a) and (b), I was struck by the perfect framing of the Orion stars while not one other star was visible. I remember standing there gazing at the constellation and feeling I was witnessing a special event that was somehow connected with the peaceful feeling I had in the tunnel under the cathedral altar. The next morning, as the bus left Santiago de Compostela that strange but special feeling was still with me. I remember looking out of the bus window at the fog-shrouded trees and trying to catch the ephemeral meaning of my experience.

It wasn't until I was writing this book – 15 years later – that the meaning of that experience was explained to me. I'll let the Angels explain in their words (Ref. Appendix 10):

> "This experience of the framing of the constellation of Orion in the skylight of the hotel room is an illustration of a very deep truth that became apparent as a result of the synchronicity. That synchronicity was given to Malcolm as an example of the way information energy can be stored in a framework. The energy of the Orion constellation was stored momentarily in the frame of the skylight as an illustration of the way that information energy of love and the life experiences of the saints are stored in the unit cells of the apatite crystals within their bones. So this love energy is stored like a battery – a battery that keeps providing love energy to those that come near. It's not necessary to make a precise physical connection to feel the benefit of that love energy. Instead, all that is necessary is to have faith and to stand very near the bones,

and that love energy will flow from the cells – it will be read out from the crystals in the bones – and be experienced by the pilgrims. Yet it does not diminish – the love energy that flows out to the pilgrims that approach close to the bones – that energy is not diminished by the reading out because it is like a library of energy. Just as you go into a library to read information from a book, you do not take the information energy out of the book, and you do not need to take the book out of the library to get the information. That is analogous to the situation of how you can read out the love energy that is stored in the bones and receive the benefits of that love energy. Yet at the same time, it is not diminished and continues in great strength for other pilgrims who can come and do the same thing time after time after time. This is one of the great gifts of God – All That Is who provides this possibility of humans receiving love energy to their great benefit, joy and strengthening of their links with Spirit."

References for Chapter 2

Ref. 2-1   *Cellulose Part 1* by E. Ott & H.M. Spurlin

Ref. 2-2   *Rand McNally Atlas of the Body and Mind*

Ref. 2-3   *Crescent Concise Encyclopedia of Science and Technology* by J.D. Yule

Ref. 2-4   *Spiritual Chemistry* by Malcolm K. Smith

Ref. 2-5   *Fingerprints of the Gods* by Graham Hancock

Chapter 3: Ghost and Spirit Energies

In this chapter:

- ❖ *Creation of a Feeling Atmosphere*
- ❖ *Lincoln Inn Ghost*
- ❖ *A Mechanism of Ghosts*
- ❖ *Spirit Energy*
- ❖ *Kirkstone Pass Inn*
- ❖ *A Spirit is an Energy Formation*
- ❖ *Main Differences Between Ghosts and Spirits*

You are probably wondering what the supernatural phenomena in the chapter title have to do with energy and crystals. Well basically, a ghost is a non-living hologram produced by the replaying of an event recorded in crystals in the surroundings. In contrast, spirits are living entities that have been in a human body. At death, the spirit leaves the body and normally returns to other dimensions, although some remain on the Earth plane for various reasons and have to absorb energy from the surroundings. The Angels of Crystal Light are going to explain the mechanisms in detail after I tell you some of my ghost and spirit stories. But first, we want to tell you about the general mechanism and how it creates a "feeling atmosphere" in some places.

Creation of a Feeling Atmosphere

I'll let the Crystal Light Angels remind you of the mechanism of energy storage in crystals (From Appendix 15):

> "We want to talk now about ghosts and spirits because that was the question you had in your mind Malcolm. You feel that there is something connecting this ghost state and psychometry – you are quite right, there is an allusion between the two. You could say that they are the positive and negative sides of the same influence. In the psychometry situation, as you know from what we have been describing, the information energy is stored in the interior of crystals. In crystals we can imagine little buckets of space formed between the atoms – what scientists call unit cells – and the information energy can be stored in the little buckets of space in the crystal. We have gone through all of this earlier, and made an analogy with musical notes stored on a three-dimensional staff where the information is stored in each measure. The notes represent little bundles of information energy that are being stored, because contained within them is information about pitch, time and all the other instructions for recreating the composer's concepts. Just as you can read out music from the measures, so you can read out the energy from the unit cells in crystals."

> "When you do psychometry, your Golden Angel (or you) goes into the crystal structure and reads out the energy, finds information that is stored there and brings it back to you. You form it into concepts and ideas about the owner of the piece that is being psychometrically read."

> "Well, a similar thing happens in a natural way in materials that have crystals embedded in their structure. As you know from your work, crystals are most commonly found in metals and minerals. The houses on Earth are mostly

built of materials like stone blocks, bricks, cement and wood, all of which have partly crystalline structures. So there is a plethora of materials suitable for storing information energy. Over long times, the events in the structures that are made of these materials get recorded in the crystals through a natural process. It's always a natural process – there is no storage of information by humans in any of the situations we have talked about in this book. It always happens through a natural mechanism where the building and the materials – and we should include textile furnishings in houses as well – all these materials absorb the information energy into the structure. If it's a very old house or a castle, then there's a background of energy existing in that place, which may be very peaceful and which has been absorbed into the crystalline materials of the building. When you walk into such a building, you feel there is peaceful energy in this place, but what is really happening is you are reading the peaceful information energy that's been accumulated over many years in the walls and the structure and textile furnishings in the building."

"But on occasions, there's some energy that is dissipated in the building or location which is far from peaceful. Situations such as the sites of big battles fought in wars or places of torture in the old days, places where murders and other dreadful deeds have been committed or other personal harm has come to some people. Energy pulses from these events go into the building materials that have crystal structures suitable for storing them, and the result is a terrific batch of energy stored in the fabric of the building. When sensitive people come into buildings where bad deeds have been done, then they are often aware of the energy that is stored in the fabric of the building. They read that energy out and understand it as an atmosphere of the place."

"You, Malcolm, have been to Culloden in Scotland, where a great battle and massacre happened. In that place, there are a lot of crystalline minerals in the ground and the energies of the terrible deeds that were done that day were absorbed into the underground minerals. Sensitive people that go to the battle site in recent years immediately become aware of the terrible atmosphere that recalls what happened there in 1746, when so many people were killed. The animals and the birds in that area are constantly aware of the bad energy in that place and they stay away from it. That is why when you went to Culloden you didn't see or hear any birds in the air or in the trees and bushes. Actually, it is a place to be avoided because of the bad energy and the birds are very sensitive to that and they stay away completely from the area of Culloden."

That deals with atmospheres that we sense in some places, and they represent a fairly low level of recorded energy. But in some situations, there has been such a terrible event that it leads to a much stronger energy storage that gives rise to what we call a ghost. Here's an example from personal memory.

Lincoln Inn Ghost

This was one of the few psychic events in my life when I actually saw something. I think I was able to see this particular one because I needed the experience for this book.

This event took place in the English city of Lincoln, where I had been staying at an old inn which had been renovated to make it into a fairly modern hotel. I was there with two colleagues while we worked at a paper mill just outside the city. Our stay at the hotel had been uneventful, and we had checked out on the last morning of our work period in anticipation of returning home. As in many paper mill jobs, something had gone wrong in the process and we were asked if we could stay for another day to get the information required.

After a day at the mill we went back to the hotel to reclaim our rooms; it was winter and we did not expect it to be full. But there was a small convention taking place there and only my colleagues were able to reclaim their previous rooms for one more night. Mine had been taken and the receptionist told me they did not have any more. When pressed, he looked at his colleague behind the reception desk and said, "Well you could have room 21." As he said that I noticed a look pass between them and I took this to mean that it was small, because I had just said I did not mind anything, even a broom closet. So I was given the key to room 21 without further comment.

Room 21 was not small, in fact it was quite large, with two single beds. The only trouble was lying on one of the beds was the body of an old woman dressed in Victorian clothes! As I stood there looking at this in amazement, the body dissolved into "wisps of smoke" which faded away over a few seconds. I guess that the whole apparition had lasted about 11 seconds.

I do not usually let these events bother me, but I was quite disturbed by this. I put my bag on the ghostless bed and went down to the bar to meet my colleagues. I wanted to tell them about it, but they were very academic personalities, and I did not fancy them laughing at me through dinner. I remember I was fairly quiet through that dinner as I wondered if I could spend a night in that room.

I decided that in the interests of psychic research I should spend the night in room 21. I had more than my usual quantity of wine with dinner to make sure I slept well, but in spite of that I slept very poorly. Although I did not see anything, I was aware of a presence. I had unpleasant dreams, and several times I awoke feeling that someone was trying to wake me up. I was glad when morning came. As we checked out we only had to leave our room keys because the rooms were on the paper mill account. I looked for someone to ask about room 21 and its reputation, but strangely there was nobody behind the reception desk.

A Mechanism of Ghosts

Paraphrasing the Crystal Light Angels' words in Appendix 15:
In most cases of atmospheres that we sense in some places, the level of recorded energy is quite low. But in some situations where there has been a very terrible stressful event (maybe in the way someone is killed with a lot of mental anguish involved), this can lead to storage of a much stronger energy. In fact, we would say it's not so much stronger energy as the sheer quantity of energy stored in the material and fabric of the place or buildings. When the bad energy reaches such high levels as it does in these places, then it doesn't wait for a human to read it out. Instead it appears spontaneously, expressing itself as a strange event that you refer to as a ghost. The energy of the ghost only comes out of the crystalline part of the fabric when human attention is directed towards it. If a human comes into a room and in the fabric of the room is stored energy of terrible events, then the terrible event energy spontaneously reads out from the crystal structure of the fabric. In modern parlance, you could say that the stored energy is suddenly triggered to download into the atmosphere. When humans witness these things, they see them as a surprising and very often frightful event or action, and they record it in their minds as a ghost. In a way, you could say the energy that is stored and appears as a ghost is so active that it "reads" itself out of the material where it's stored – it downloads itself from the fabric of the building into the atmosphere and mental processes of the human who is aware of something strange happening.

That is why in the Lincoln Inn there was a ghost in room 21. In that case, the person that gave rise to the intense vibration stored in the fabric of the walls – stone walls by the way, so there was a lot of crystalline material available to record the events – that person was jilted in love. She was living in the building which was an inn in those days too. From the clothes of the ghost, you saw she lived in Victorian times (1837–1901). When you walked into the room, the emanations of energy from the walls had molded themselves into the form the woman took as she lay on her deathbed. She was so unhappy, so completely awash in anguish, that she took poison to kill herself, then she lay on the bed to die. As she died, her energy was stored in the fabric of the walls of that particular room. In later times when anybody went in that room, the stored energy automatically downloaded a reading of itself.

It's like having material in a computer memory and every time you want to see that material you can print it out and take away a copy, but it doesn't change the stored information at all. It remains in the computer and is not decreased in amount or intensity, and at any time you are able to read it out as a printed sheet of paper. In all these cases where spontaneous appearances occur, it's like the computer deciding it's going to print out a page or two of text because it's so important to the originator of the program. It seems that the originator leaves instructions that every time a human pays attention to the program, it should download another print and make an appearance in apparently physical form. The woman you saw lying on the bed as you came in was not made of physical flesh and bones, but was like a hologram of energy and that hologram had been stored in the walls of the building.

Basically that is the mechanism of producing a ghost, which is energy recorded in the surroundings – that is why the ghost's actions are usually repetitive. We can tell you that all the time there is no human around, everything is peaceful, but as soon as the stored energy detects the presence of a human that can receive the message – which is to hear screams or chains rattling or see a vision of a person maybe with a physical deformity relating to the way they died – as soon as the human is detected by the energy, then it automatically downloads the energy performance that we see as a ghost or hear as ghostly noises.

Spirit Energy

The first time I experienced the energy loss from a house to an Earthbound spirit was when my mother died in England in 1974, just after I had emigrated to Canada. My cousin Irene told me as she took me to my Mum's house that she believed my mother's spirit was still in the house because it was so cold, although the weather was not. Irene was right – when I returned alone later that day, Mum's spirit talked to me and we had a little ceremony that released her. I felt her go and the house temperature returned to normal quickly.

Kirkstone Pass Inn

(I understand if you have the impression that I spent all my time in England in inns and pubs – but there are so many!)

In the 1990s my wife and I were on a bus tour of England. One evening, we went on a dinner excursion to an 18th century coaching inn located high on the Kirkstone Pass in the English Lake District. After dinner we were invited, by the young couple who owned and ran the inn, to go up to one of the rooms which they said was haunted by a coachman they called Neville. They had slept in one of the guest bedrooms and had been awakened one night to see a shadowy figure dressed in the heavy coat of an 18th century coachman. It seemed the name Neville had been passed down by previous owners of the inn, who were also familiar with him.

Of course, everyone in the party wanted to see the room. Since they had consumed some of the other kind of spirits during the meal, there was a stampede into the tiny room with everyone laughing and joking and calling out to Neville. Apparently nobody in the party perceived anything unusual, because most of them left the room and laughed all the way downstairs.

When the crowd had gone, I entered the room and was immediately struck by the cold atmosphere in there, which was amazing since about 20 people had just been milling around in it. I walked around and found a sort of alcove where the cold was intense. It was like being in a walk-in refrigerator – I have been in meat storage lockers that felt like that. It seemed that all the heat was being sucked out of my body.

The lady owner was waiting to shut up the room and she watched me with a mysterious smile. I asked her, "Is it always as cold as this?" She laughed and told me I was the first in this group to notice that. She went on to relate how it was the intense cold that awoke her and her husband on the one night they saw Neville. He stood in the alcove, which I had found to be the source of the heat loss, and which used to be a doorway to the mountain outside. She told me that they found it so cold they could not sleep in there again in spite of putting extra heaters in the room.

I did not have a thermometer to measure the temperature, which would have been very useful to answer one of two questions:

- Is the area really lower in temperature or is it that some people are able to sense the energy being drained by the presence of a spirit?
- Are all people aware of this coldness or just those with psychic sensitivity? (In the case of Neville's spirit, why did none of the people who visited the bedroom sense the coldness that seemed so obvious to me?)

A Spirit is an Energy Formation

The Angels explain about spirit energy in Appendix 15:

"A ghost is energy that is recorded in the surroundings.

That is quite distinct and different to the appearance of a spirit, which is an energy formation that was once in a human body – or an animal body for that matter. It exists in its own right. Even in heaven, before a person is born there is an energy construct that we recognize as a spirit. We are all spirit beings. You humans sometimes take on the physical form of a human to experience and learn. When that experience – your lesson – is over you return back to the place you call heaven, the spirit world or the 5th dimension. You are not recorded in any situation like a ghost but you could say your spirit is recorded in the body that you build."

"When a spirit is released from the body at death or in an out-of-body experience it can travel as an energy construct. In certain circumstances, those energy constructs that you call spirits can make appearances to people who are sensitive and they can communicate with a spirit that is in the body – that of course is what you call a human medium. Those spirits are quite distinct from ghosts that we were talking about earlier. Spirits have to absorb energy from the surroundings to give them energy to move and exist when they are on the Earth plane. Just as you have eaten your breakfast today, Malcolm, to feed your body, you feed your spirit with love. All energy is love energy really. When a spirit is trapped in a place, for example, your mother still in her house waiting for you to return from Canada, then to exist, to have energy acting as nutrition, that spirit pulls energy from the surroundings in the building where it is staying. When a human comes into that building it detects the energy loss as a cold atmosphere. There is an actual decrease in the temperature of the building it's not just a physical effect on the body of the perceiving human."

"If you look in the book *The Haunting of Borely Rectory* (Ref. 3-1), you will see there was a measurable decrease in temperature. In that book a ghost-hunting group measured the temperature differences at several degrees Fahrenheit as a measure of the amount of energy that was absorbed by the spirit trapped in that location."

"The most intense reduction of temperature resulting from spirit absorption of energy was the one you experienced in the haunted inn in the English Lake District. You know which one we mean, Malcolm. You felt like you were standing in a walk-in refrigerator when you stood in the alcove that had once been the doorway. That was because the spirit that was still trapped there, since coaching times, had needed energy to maintain itself so it subtracted heat energy from the building resulting in very low temperatures in one particular room. If you encounter such spirits, then it's your duty as a human, being of the spirit world, to try to help the trapped spirit be released. You helped the spirit in the Lake District Inn by praying for its release and telling it that it could go on to live in the next dimension, as all spirits do after death. In that case you were partly successful, but other people had to add their assistance in later years since you were there, so it didn't happen right away."

Main Differences Between Spirits and Ghosts (from Appendix 15)

"The energy that is perceived as a ghost is stored in crystalline material in the fabric of the building, or in the ground under the event, and that energy is read out automatically. It's like printing out a page from a computer memory since the amount of energy that is stored is not decreased by each reading out. When the energy detects the attention of a nearby human, it downloads into the atmosphere and the human perceives the energy as a ghostly form or noise."

"A spirit is an energy formation that was once in a human body – or an animal body for that matter – it exists in its own right. Even in heaven, before a person is born there is an energy construct that we recognize as a spirit. We are all spirit beings. You humans sometimes take on the physical form of a human to experience and learn. When that experience – your lesson – is over you return back to the place you call heaven, the spirit world or the 5th dimension. You are not recorded in any situation like a ghost but you could say your spirit is recorded in the body that you build."

"In the case of a spirit that is trapped on the Earth plane, maybe as a result of its own need to settle something that happened during its own lifetime, it needs energy to maintain itself in the Earth's atmosphere. So it removes heat energy from its surroundings and this is perceived by humans as a deathly chill in the building."

Reference for Chapter 3

Ref. 3-1   *The Haunting of Borely Rectory* by Eric J. Dingwall, Kathleen M. Goldney and Trevor H. Hall

Chapter 4: Storing Energy in Liquid Crystals

In this chapter:

- ❖ *Liquid Crystals*
- ❖ *Lipid Double Layer*
- ❖ *Cell Wall Structure*
- ❖ *Storage of Information by the Lipid Double Layer*
- ❖ *Transmission of the Information to New Generations of Cells*
- ❖ *Stored Information Accompanies Organ Transplants*
- ❖ *Transfer of Acquired Information in the Animal World*
- ❖ *In Conclusion*

So far I have talked about storage of information/energy in:

- Regular solid crystalline materials such as minerals and metals.

- Crystals (usually very small) in semi-crystalline materials such as wood and textiles.

Now I am going to describe storage of energy in liquid crystals, which are familiar to us as the active component in many electronic displays. You may be surprised to learn that liquid crystals have an important part to play in the cells that make up our bodies – far more important for humans than the electronic applications.

Liquid Crystals

A scientific definition of liquid crystal is a substance that flows like a liquid, but has some order in its arrangement of molecules (Ref. 4-1). In the order is the structure that acts like a crystal.

First we have to look at the shape of the molecules that form liquid crystals. Most molecules form liquids that have random structures. Remember my description in Chapter 1 of glass being a structureless liquid that cooled and solidified too fast for the molecules to line up "on parade?" With liquid crystal molecules, their shape and chemical properties help the molecules line up on parade.

Lipid Double Layer

Most of the molecules that form liquid crystals are shaped like lollipops with two sticks. Typical are the phospholipids – it's okay you don't have to remember that name – I'll call them just lipids for short. They are the fatty materials that form animal cell membranes. These molecules are made up of a phosphate group that acts as the head of the lollipop, and two hydrocarbon groups which form the sticks of the lollipop (Fig. 4-1). There's an important chemical difference between the head and stick parts of the molecule. The head prefers to be in water – chemists call that "polar," because some of the atoms are electrically charged – and the sticks, being hydrocarbon with no charge, prefer to be in oil.

When you put such molecules in water the polar heads eagerly associate with the water molecules. But the hydrocarbon stick parts want to get away from the water, so they cluster in a double layer where their exposure to water molecules is minimized. A diagram of the resulting lipid double layer is shown in Fig. 4-1. So as a result of the chemical forces between the molecules, they space themselves apart uniformly like "soldiers on parade." In this ordered state, the molecules can be said to form a crystal.

The fact that these crystals are liquid means that they can flow, but in doing so they keep their regular spacing no matter how the liquid bends as it flows. Coming back to the "soldiers on parade" analogy, we can say that the soldiers march, keeping their uniform spacing between themselves – even if they turn a corner! They are behaving like a flowing liquid yet they do not lose their "crystalline" organization.

In his book, *Biology of Belief* (Ref. 4-2), Bruce Lipton explains how cell wall membranes are made from a lipid double layer. Their fluid crystalline organization allows the cell wall to change its shape while maintaining its integrity.

Cells have made use of the double layer to create a defensive wall around their cytoplasm (stuff inside the cell) which is fragile chemically. If we looked through a microscope at a cross-section – slice – through a cell wall we would see something like the diagram Fig. 4-2. The dark spots in the double layer are the phosphate heads of the phospholipid molecules. Between the two dark lines is a pale layer, and that is the hydrocarbon oil-loving part of the double layer. This important oily lipid layer is a barrier that prevents positively or negatively charged polar molecules from crossing it. In effect, the lipid layer acts as an electrical insulator that keeps polar molecules in or out of the cell.

In actual fact the cells need some charged molecules as nutrients. These needed charged molecules are allowed through the lipid layer by special proteins (Integral Membrane Proteins) that reside in the double layer and act as gates. You can read about these gates in Bruce Lipton's book (Ref. 4-2), where their structure and function are described clearly. But for our purposes we do not need a detailed description of how nutrients and waste products are moved through the barrier. All we need to know is that there is a phospholipid double layer around the outside of every living human cell and that it has the structure of a liquid crystal.

Storage of Information by the Lipid Double Layer

The Angels of Crystal Light describe how the liquid crystals store information energy (Appendix 11):

> "We want to tell you today how that lipid double layer is used to contain information. You have seen it is very small, but of course it has to be small because the cells are small. The trillions of cells that make up your bodies contain the double layer of lipid molecules, which is used as a chemical defense against the environment in which the cells find themselves. They contain within their wall all the things necessary for life in the cell, in particular the nucleus where the DNA is stored. These cells have a wall that goes around them in all three dimensions and in that wall is the double layer of lipids. In that double layer is what humans have called liquid crystal because it is partly in a liquid state, and it is relatively mobile so that the cell can change its shape. But it is nevertheless a crystal because the molecules are aligned. This alignment of the molecules results in crystal unit cells which are quite flexible, but usually maintain a shape which approximates a cubic type of crystal unit cell."

(Your attention please Dear Reader: Science uses the word "cell" in two different ways: 1. A unit of living matter. 2. The smallest unit of a crystal's structure formed by atoms or molecules – see Figure 1-2. I have called the former a "cell" and the latter a "crystal unit cell.")

Continuing with the Angels' words:

> "But it is not strictly adhered to, because the cell needs to move and the liquid crystal needs to flow a little to accommodate the cell wall movement, so the shape of the crystal unit cell changes slightly. Nevertheless it is still an efficient holder of information energy. That informa-

tion energy is stored in the liquid crystal just as in regular solid crystal."

"Those flexible crystal unit cells are just as capable of storing energy as those unit cells in a crystal of aluminum sulfate for example. Just because the lipid double layers are fluid and flexible doesn't mean to say the atomic forces in the liquid crystal are different than in a solid crystal. Those atomic forces are quite capable of holding information and then passing it on from generation to generation through changes in the DNA of the cell. So it is a beautiful mechanism that was created by God, and it works very well and unobtrusively. It is so unobtrusive that your scientists have not fully understood it yet, in fact most of them hardly understand it at all."

Transmission of the Information to New Generations of Cells

"We can see that you are concerned how the information can be perpetuated when the cells keep changing as they divide to create new generations of cells. You are right in thinking this, but once a cell has acquired a unit of information energy – a byte you could say in modern computing terms – that byte becomes part of the cellular structure, part of the lipid double layer in the cellular wall, and that information is duplicated in any daughter cells that come from the parent cell. There are instructions in the DNA that tell the cell to duplicate any information that is held in the double lipid layer. This is something that is hard to explain in physical terms. Let us say that the parent cell has intent as a result of the information that is stored in its cellular wall structure. That intent is held by the DNA in the cell nucleus, so that when the cell replicates itself before its death, it does so not only replicating the physical layout of the cell structure but also the information that is there."

"This system works well and transfers from generation to generation not only the basic information of the cell but also its acquired information. In this respect it is quite 'Lamarckian' – as you know, Lamarck was a Russian researcher who said that cells pass on from generation to generation acquired characteristics. This is a prime example of Mr. Lamarck's vision in seeing that cells would pass on information that had been acquired during the lifetime of the cell. Here we have a way of acquiring information and storing it in the liquid crystalline structure of the cell wall, and we can be sure that the information will be duplicated as each generation of cells come forward, because the acquired information is passed from each generation of cells to the next cell generation. This is a beautiful system that works well to acquire, retain and pass on the acquired knowledge and habits of any organism."

"It's a part of being the multi-cellular organism that humans are. Every cell in the human body has to die and be replaced, and if there were no way of passing on the information acquired by one generation of cells, then there would be no learning for the whole organism. So it is necessary to have this process by which not only the structure of the cells is passed on and replicated, but also the acquired information – if you like the memories of the cells – is passed on through the DNA memory retention system, which is quite complex. As you know, there are twelve layers of DNA, and in some of those layers is the mechanism which ensures that the cell memories are passed on from generation to generation of the cells. This is a beautiful mechanism that works without any trouble at all. But because human scientists are so certain that memories are associated with the brain, they do not see that every cell in the body of a human – or animal – has its own experience and memories and passes them on to subsequent generations of cells. It's very necessary that this should happen because some cells live in the body for only a few days, or a few hours in some cases. So it's very necessary that the cells not only have a memory – a cellular memory as Kryon calls it (Ref.4-3) – but also there is a mechanism for passing on the memories, so the new generation of cells learn from the experiences of the older generation of cells."

"It's a bit like humans. One generation of humans adds to the educational system of the upcoming generation. You pass on the information to the generation that is taking over the world and in the same way each cellular generation passes its information on to the next generation of cells that is to form the bodies of humans. It's a learning process that never stops because each generation receives the benefit of the previous generations' experiences. That's why Mr. Lamarck was quite correct in his beliefs that acquired characteristics were inherited, although a lot of people doubted him at the time because they could not see a mechanism. More and more scientists, biologists in particular, are now seeing the wisdom of Lamarck's perception that this must be so to explain how the specialized behavior of certain species can be developed. It's no good learning by experience if the learning of one generation is not passed on to the next generation, whether it be humans or cells. It is the same process working in each case." (Appendix 11)

Stored Information Accompanies Organ Transplants

"You have heard about organ transplants, in particular heart transplants, that pass on to the organ recipient life characteristics of the donor. For example, in *Biology of Belief* (Ref. 4-1), Bruce Lipton recounts the story of a health-conscious young woman who developed a taste for beer, chicken nuggets and motorcycles after she received a heart-lung transplant. Her research showed she had received the heart of an 18-year-old motorcycle enthusiast who loved chicken nuggets and beer. That was simply because of the information that was stored in the heart cells, in the crystal structure of the heart cell walls. This information in the donor's cells was passed on to the woman now that she was using the heart that he had grown. Not only the heart, but all his life habits and preferences, were passed on as well in the walls of the heart cells. So this is a very powerful memory tool that can provide surprising information."

"Not only does a heart cell need to know about how to operate as a heart cell – the contractions of a heart go on for years and years – but also the heart cells need to acquire a knowledge of the whole organism's experience and to hold that in their cellular memory. They pass on that knowledge to the next generation of heart cells, which benefit from the experience, so that acquired characteristics are passed on. That is why if a heart is taken out of a donor body and put into a recipient body, it still carries the memories of the donor heart cells which get passed on to the recipient. Sometimes the recipient acquires surprising habits or characteristics, as in the case of the beer and motorcycle lady we told you about earlier. But it all makes good sense that it should be like this, and this is a mechanism that has been created by God and it works very well without any interference from other sources, because the acquired characteristics are locked away in the cellular memories. Even if a section of those cells is lost, the memories from the cells transfer to the DNA, and pass from human generation to generation through the transfer of DNA that occurs at the conception of a child." (Appendix 11)

Transfer of Acquired Information in the Animal World

"We know from Kryon's description (Ref. 4-3) that the cellular memories are passed down from human generation to generation. In addition, cellular memories make it possible for animals and birds to acquire amazing specific memory patterns such as bird calls and habits, like feeding on platforms containing food that humans have put out. The latter habits have been acquired in recent times, since there were no situations like this before. Some birds acquire special habits; we are reminded of the blue tits, those little birds in England that break into aluminum foil caps on bottles of milk left on door steps so they can drink from the milk. The ability to make a hole in aluminum caps to get a drink of milk is an acquired characteristic that is passed on to the new generation. As soon as a bird is hatched from its egg it knows it can get that kind of food in that particular way. How do you think that knowledge is transferred from one generation of blue tits to the next? It's through cellular memories such as we have been describing." (Appendix 11)

In Conclusion

"It's a very powerful learning tool for people, because the human race would not have learned so quickly if it had to depend on teaching. Habits of behavior that it had acquired over generations through teaching from one generation to another are important, of course, because the brain is involved in that activity. But at the same time, there are acquired characteristics that are not taught but which are passed on from generation to generation. Music is a case in point here. We know music is taught in some schools, but many races have an acquired love and knowledge of music that is passed on through the memories of the cells in their bodies. So you could say that music is in their blood and that would be very true – literally."

"We are sure you realize this book writing that we do together is a combined operation of different levels, which in one way reflects the different levels of memory. There is brain memory and cellular memory. You could say that there is an analogy for our relationship, in that we Angels represent the brain memory and you, Malcolm, represent the cellular memory. It's a rough analogy, but it could be used to describe different levels of memory and how they operate among the cells of the human body." (Appendix 11)

References for Chapter 4

Ref. 4-1   *A Dictionary of Chemistry,* Edited by John Daintith, Oxford University Press

Ref. 4-2   *The Biology of Belief* by Bruce H. Lipton, Ph.D.

Ref. 4-3   *Kryon Book 7 – Letters From Home* by Lee Carroll

Chapter 5: Cosmic Lattice Psychometry

In this chapter:

- ❖ *Introduction*
- ❖ *Summary of the Techniques*
- ❖ *The Cosmic Lattice*
- ❖ *Connecting with the Cosmic Lattice*
- ❖ *Psychometric Techniques – The Flower Method*
- ❖ *Psychometric Techniques – Sand Impressions and Coloured Ribbons*
- ❖ *Psychometric Techniques – Past and Future*

Introduction

There are at least three techniques of psychometry that do not rely on reading energy information from crystalline materials. Instead the information is drawn from relevant parts of the cosmic lattice. These three techniques are usually referred to as:

- The flower (or plant) method
- Impressions in sand
- Coloured ribbon arrangements

Probably there are many other similar procedures in other cultures since this type of divination is very old. My recent experience has been limited to these three procedures and the Angels of Crystal Light have offered mechanisms only for these, therefore in this chapter the discussion will be limited to these three.

Summary of the techniques

In each method the person who wants a reading – for convenience I have called that person the "readee"– either:

- brings a flower to the session,
- makes an impression of both his or her hands in a tray of sand,
- arranges a few – typically four or six – coloured ribbons in order of preference of the colours.

The psychometric reader then studies the offerings and is led to make pronouncements about the readee's life and or personal characteristics.

In training and practice sessions, the identity of the readee is unknown to the reader, who has only the physical offering to work on. In my personal experience, readers can reveal an amazing amount of personal detail, which the readee acknowledges as correct when the time comes for their identity to be revealed.

The Cosmic Lattice

Since the Crystal Light Angels have told me that these psychometric techniques all involve the cosmic lattice, let us review the information about it that was presented in our first book (Ref. 5-1).

The cosmic lattice is built and maintained by hyper-dimensional love energy that comes to us through Sirius, our sun and planets. The energy comes out of the planet surfaces at the touching points of the inscribed double tetrahedrons, which probably represent standing energy waves inside each planet. Some of it doubles back above the planet surface forming hexagonal cells in an energy matrix.

NASA has photographed a hexagonal "cloud" above Saturn's north pole (Ref. 5-2). I am told by my Guides:

We confirm that the hexagonal cloud seen above Saturn is part of that planet's cosmic lattice segment.

We point out that this hexagonal cell is only part of the infinite cosmic lattice.

The cosmic lattice is around and through every known body of matter in the universe.

As my Guides have stated, the cosmic lattice is everywhere in the universe: it passes through all matter. I would like to add to their definitions the following points about it that I have learned from the Kryon writings (Ref. 5-3):

- Throughout the universe is an invisible energy matrix called the cosmic lattice.
- The actual cosmic lattice exists simultaneously in all 12 dimensions of the universe.
- Its 12-sided honeycomb cells are closed like human body cells. The cells lie side by side but do not touch each other.
- The cosmic lattice is the energy source of the universe; it is the energy of Spirit; it is the energy of Love.
- It responds to human consciousness; it is the basis of telepathy. Humans and animals intuit what is happening far away by tuning in to waves on the lattice.
- Communications are virtually instantaneous, in comparison light is slow.
- The cosmic lattice is the basis of synchronicity; we use its energy to create our futures; it is the very essence of inspiration.

My Guides tell me that all the information we need to create anything (paintings, music, poetry, inventions, futures) is stored on the lattice. So are the morphic fields for all the animals and plants and crystals and molecules in the world as described by Rupert Sheldrake (Ref. 5-4). For example, a cutting of a willow tree has within its atoms the cosmic lattice. Through that it has a connection to the morphic field of all willow trees, which is part of the lattice. It is this morphic field that tells the willow cutting how to grow into a beautiful tree. In fact, all morphic fields, templates or designs that exist as energy are part of the lattice. They are woven together in a way that we on Earth would call holograms.

A hologram is a three-dimensional image, made of light, that is stored on photographic film. The image is reconstructed by shining laser light through the photographic film. Information about any part of the image is stored ALL OVER the photographic film. In fact, if you cut off the corner of the film, there's still enough information left there to recreate the whole image, all that is lost is a little definition of the image. You can even create an image from a small fraction of the original film, although the picture isn't very clear. The energy fields that are part of the cosmic lattice are spread over the cosmic lattice hologram throughout the universe so they are accessible from any part of it.

Connecting with the Cosmic Lattice

I will let the Angels tell you how we connect with the cosmic lattice:

> "Most humans use the cosmic lattice unconsciously – we've talked about this before. In the unconscious use of the cosmic lattice it seems you just give attention to things, good or bad, and you get whatever you think about. That is the basic unconscious use of the cosmic lattice that most humans practice without being aware of it because it's unconscious. But when a human needs

to make contact, and does so consciously, then he or she enlists the assistance of his or her Golden Angel (see Chapter 1). That is the way the connection is made. This is because it is difficult for humans to make the connection consciously, although they can learn to do it. Many people who are proficient at telepathic communications are able to make the actual connection with the cosmic lattice themselves with or without the assistance of their Golden Angel." (Appendix 12)

Psychometric Techniques – The Flower Method

The Angels continue from Appendix 12:

"Let us choose the flower method to talk about in detail first. With this method, the flower is connected into the cosmic lattice, which holds the flower's growth patterns, the flower is in communication with all other flowers and with all other living things on the cosmic lattice. As you know, it connects all living things which include rocks and things you probably would not think of as living, but nevertheless they are. All the living things are connected into the cosmic lattice because they helped create it for this planet. Each particular species of plant has a particular part of the cosmic lattice."

"The flower used for assistance in psychometry has a connection to the cosmic lattice and that's what the Golden Angel uses to find the information required by the human reader. For example in the psychometry lessons which you Malcolm talked about in the introduction to this book, Hazel brought a piece of bamboo, which was the plant that you chose to work with – although you did not know that it was Hazel who had brought it. That bamboo plant led you to connect with your Golden Angel Mikael, who entered the cosmic lattice via the plant's connection with the part of the cosmic lattice that deals with the affairs of bamboo plants. When Mikael and the bamboo part of the cosmic lattice were connected, Mikael was able to use that connection to go into another part of the cosmic lattice that deals with Hazel – the 'owner' of that particular piece of bamboo – and Hazel's life and Hazel's likes and dislikes. (My Guides tell me that there was an overlap between the bamboo part of the lattice and the part related to Hazel as owner of that particular piece of bamboo. Mikael was able to detect the overlap and so made the connection between Hazel and the bamboo. – MKS) One of the things that Mikael found out about Hazel was that she liked aromatherapy, so Mikael brought that back to you, Malcolm, and you perceived the concept of aromatherapy in connection with Hazel. You announced to the class that the owner of the bamboo liked aromatherapy – among other characteristics – and Hazel revealed that she had brought the bamboo and she really liked aromatherapy."

"And that's basically how it works, the plant provides a way or a route into the cosmic lattice. From there, the lattice can be explored fully and connections made with the apparent owner of the plant – we know Hazel had bought the bamboo only a few hours before the reading, nevertheless, it stood in for her as a way of connecting to her special part of the cosmic lattice. There, information concerning Hazel was stored, and Mikael could select some suitable piece to bring back to you, Malcolm, so that you could do a psychometric reading. In that case, it worked for a person, but of course you can get information on many factual areas of the cosmic lattice by using a plant as an entrée into the lattice."

"The cosmic lattice is accessible to all the Golden Angels. They are quite familiar with exploring and getting information from it. They don't really have to explore it because they know just by thinking about it where all the different parts are. As we told you before, if you wanted to know something about strawberry jam you only have to think of it, and if you are an Angel, you go immediately to the part of the cosmic lattice concerned with strawberry jam. If you are a human, it may take a while for you to find it because you are not practiced at knowing all these things. If you get the help of your Golden Angel, then he can show you the way."

Psychometric Techniques – Sand Impressions and Coloured Ribbons

Description of the mechanisms by the Angels from Appendix 12:

"There are several other psychometry techniques similar to the plant method. There's the sand technique, in which people push their hands into a bed of sand to make an impression. That is another route that the Golden Angel can follow to make the necessary connection. There is another technique using a bundle of coloured ribbons that are attached to a frame at one end. The person who is to have the reading – we call that person the 'readee' – puts the ribbons in an arrangement of colours which is pleasing to them. In arranging the ribbons or making a sand impression, a personal energy stamp is made. You have heard of energy stamps. Well, in this technique, the readee puts a personal energy stamp on the arrangement of colours of the ribbons, putting on top the preferred colour, putting underneath that the second preferred colour and so on. The arrangement of colours signifies to the Golden Angel of the reader an energy stamp relating to the personality of the readee. In this case, and that of the sand impression, the Golden Angel does not need to go through a natural object to get to the area of the cosmic lattice relating to the person being read; instead, the Golden Angel goes directly to the lattice area relating to the readee, because that person has established a personal energy stamp which itself acts as a route for the Golden Angel to connect directly with the lattice area concerned with the readee's life. So that is a slightly different technique. All these methods are working for the same purpose and that is for the Golden Angel of the reader to make a connection with the readee's personal part of the cosmic lattice."

"To summarize we can say the purpose of all these indirect methods, such as reading information about the owner of a plant, an impression in the sand or arrangement of ribbons, is to connect with the special part of the cosmic lattice which is concerned with the readee's life and preferences. One method is to use a plant where the route into the cosmic lattice is that plant's special part of the lattice. Another method is to go directly to the readee's special part of the lattice. In that case, the readee establishes a personal energy stamp by sand impressions or arranging colours of ribbons."

Psychometric Techniques – Past and Future

Paraphrasing the Angels' words from Appendix 12:

So those are quite efficient methods that have worked for thousands and thousands of years on the planet Earth. "People who were viewed as seers used these methods in a natural way and weren't interested to know how these things worked, because in those days people did not think about things scientifically. Even when they did start to think scientifically it was about physical objects – for example, why cannon balls of different sizes all fall at the same rate. It wasn't until Descartes wrote 'I think therefore I am' that any questions were asked about humanity's methods of connecting with each other or with the universe at large. That's when people started to become curious to know how these things work, especially people like you, Malcolm. You were born curious and that was set up deliberately because, as you were told by your Guides, you were trained by us before you came into this lifetime to be a very curious, questioning person, because you felt you had to understand God's plan for the universe. That is a very worthy attitude to have and it brings you to the point where you can find out many things, explain them to your fellow humans through the books and do a great service. We realize that there are not many people who are curious about these things, because so many come into the Earth plane with curiosity and a wish to understand the Earth, but they lose their train of thought and very quickly become totally immersed in physical things. That is a pity, but things are changing in that respect."

"Now there are the first instances of scientists realizing that it's important to take into account God, His creation of the universe and the love that flows throughout that wonderful universe. Without understanding of that love, scientists cannot pursue their investigations that are concerned with only physical things. It is not possible to explain the universe unless you take into account God and the love He has given the universe. That love is the essential driving force of the whole universe as humans perceive it. Without love it is not explainable. But you have made a good start in our first book in putting that value of love forward as an important scientific 'variable,' if you like. You said in the introduction to our first book that 'Science doesn't make sense without God.' We hope that as a result of the books we are going to write, that lesson will come through and scientists, and all people on Earth, will understand that the universe and its structure, its meaning and its purpose are only explainable in terms of God, All That Is and the wonderful love He showers upon us all."

References for Chapter 5

Ref. 5-1 *Spiritual Chemistry* by Malcolm K. Smith

Ref. 5-2 *The Monuments of Mars* by Richard C. Hoagland

Ref. 5-3 *Kryon Book 7 – Letters from Home* by Lee Carroll

Ref. 5-4 *The Presence of the Past* by Rupert Sheldrake

Chapter 6: Homeopathy and Crop Formations

In this chapter:

- *Introduction to Water Chemistry*
- *Dr. Emoto and the Hexamers*
- *My Musical Water Experiments*
- *Water as a Recording Medium*
- *Homeopathy and the Crop Formation Analogy*
- *Water as a Recording Medium for Healing Love Energy*
- *Persons Interfering with the Homeopathic Process*
- *More on Solid Homeopathic Remedies and Mechanisms*

Introduction

In this chapter I'm going to tell you about storage of information energy in water – that's right, plain old-fashioned liquid water. It's not even frozen water so there are no (ice) crystals to hold the information in their unit cell "energy buckets" as there were in the first three chapters of this book. Water is amazing stuff – perhaps that's why we've been given so much of it on our planet. First I need to tell you about structures that water molecules like to settle into.

Introduction to Water Chemistry

For those of you new to chemistry, let me remind you of some concepts. The basic building blocks of our world – and the whole universe as far as we know – are atoms. They consist of tiny bundles of energy with names like protons, neutrons and electrons. The protons and neutrons stay in the central nucleus and the electrons buzz around the nucleus in orbits like a very, very small solar system. In our world, there are about one hundred different kinds of atoms. Each kind of atom is called an element – things like sulfur, iron, hydrogen and oxygen.

If hydrogen is burned in an oxygen atmosphere, two atoms of hydrogen combine with one atom of oxygen to form one molecule of water. That's why the chemical formula is H_2O. Please notice this defines a molecule as a little group of atoms bonded together. But not all molecules are so small, for example, the protein molecules that make up much of our bodies are big enough to be seen with a powerful electron microscope.

The "glue" that sticks atoms together to form molecules is electrons. In the water molecule, the oxygen atom has got two spare electrons that are looking for something to "stick" to. So, in almost every sample of water, the oxygen atoms stick to hydrogen atoms in adjacent water molecules – this is what chemists call hydrogen bonding. It's only a weak bond, but it's responsible for water molecules sticking together in clusters called oligomers.

Dr. Emoto and the Hexamers

I know this section heading sounds like the name of a rock group, but that kind of music is not what is needed for the most beneficial structure of water, as we will see when I get into some of the experiments I've done in that area. Before I go any further, I must explain that in addition to oligomers, there are also hexamers, clusters of SIX water molecules. Because the bond angles work out just right, the six water molecules can form a ring – really a hexagon held together by hydrogen bonds – and the ring structure makes the hexamers more stable than other size clusters. Nature takes advantage of this fact, as we will see later.

Now from Appendix 13 the Angels of Crystal Light introduce to you Dr. Emoto's work (Ref. 6-1):

"As you know from Dr. Emoto's experiments, water can be influenced by energy patterns and human thought. Normally water molecules exist in clusters called oligomers which are bonded together by hydrogen bonds. Water oligomers occur in a range of sizes consisting of a few to many water molecules depending on the natural conditions of the water. Spring water is more beneficial for humans than stagnant water because in rapidly moving natural water the oligomers are relatively small (about six molecules). If the water is stagnant, and especially if it becomes polluted with negative energy from either human or natural sources, then the water molecules are slower moving because the energy is low and they tend to collect together into relatively large oligomers consisting of up to thousands of molecules. However, under the influence of human intent, which is expressed as good or happy thoughts, good wishes and prayer – any of those forms of human thought – the oligomers are constrained into smaller groups of the order of six molecules. As you know from your structural chemistry, six molecules bonded together as hexamers are the most stable configuration, that is why snowflakes have six-fold symmetry."

"Most of Dr. Emoto's work was based on the perfection of ice crystals that were formed from water from different sources. Perfect ice crystals were formed from 'good' water – that is either fresh spring water or water that has been blessed by humans – so that the water consists mostly of hexamers (six molecule clusters). The beautiful ice crystals from good water mostly had the same structure as natural snow flakes with their six-fold symmetry. Ice formed from polluted or stagnant water appeared to have a non-crystalline structure that resembled congealed mud."

"The stable hexamer structure of water provides the most efficient use of water required by plant and animal life on this planet. Water is the liquid of life on Earth. Your human bodies consist of at least 70 percent of water depending on which part of the body you are talking about. The average is about 75 percent. The water in human, animal and plant 'bodies' has to flow through very small orifices in cell walls and if the water is in very large oligomers, its apparent viscosity (thickness) is too great for efficient flow in and out of the cells. If you look up the dimensions of orifices in the cell walls, you will see that they are slightly larger than a water hexamer, so that hexamers can slip through the orifice easily, the apparent viscosity is low and water flows through the cell efficiently. On the other hand, if the water is present as oligomers consisting of about 10 or more molecules, then they tend to get stuck in the orifices. If that happens, the apparent viscosity of the water is high and the flow of water through the cell is slow and inefficient."

"Water in the form of six-molecule clusters called hexamers fits through the cell wall orifices easily and good flow conditions are experienced by the human, animal or plant cells. Under these conditions, the water is experienced as beneficial to the life of the cells and the whole organism thrives on the hexamer-based water, because it flows through the cells quite rapidly and everything works efficiently. When stagnant water is used, and that includes water that has been exposed to negative influences, then the water is in relatively large oligomers. This situation leads to stagnation of the water in the cells, because the large oligomers become tangled and take a longer time to move through the orifices. The apparent viscosity of the water is increased. Multicellular organisms like humans, animals and plants don't thrive on water containing relatively large oligomers. (As an aside, we can say that if you send love to your plants they will thrive because the love that you send equals good intent which is part of the love energy that we talked about in our first book. [Ref. 6-2] Love energy puts the water molecules, which are already in the bodies of the plants, into the most stable and beneficial form, which is hexamers – or smaller clusters – throughout the plant's structure. The water flows through the plants very readily without any blockages and the plants are 'happy' and grow strong and beautiful.)"

My Musical Water Experiments

I used to work as a chemist in the papermaking industry and my experience in studying how water soaked into paper suggested a wicking experiment that might detect a difference between water treated with "love" and "hate." Those were the words I chose to put on labels, which I put on two identical new glass jars, holding about 1 litre each. The jars were filled with Vancouver tap water sampled at the same time.

(I will spare you from having to read all the technical details – I've put those in Appendix 16 at the back of the book – but I have to record the details of my experimental methods if anyone from conventional science is going to believe the results I obtained.)

The jar labeled "love" was put in my car trunk overnight where it "listened" to classical baroque music for 18 hours. The next night the jar labeled "hate" was exposed to discordant rock music for 18 hours in the same car trunk. (During the preparation of the samples, the car was in a closed garage on consecutive mild nights.)

For the wicking part of the experiment, coffee filter paper was cut into strips (1 centimetre by 10 centimetres), a line of water soluble ink was drawn across the strip near the bottom and the bottom 0.5 centimetres was dipped and held in a small sample of the 'love' or 'hate' water. The water was drawn up the vertical paper strip by capillary attraction and the water picked up the soluble ink as it passed the ink line. The ink served as a marker for the water front as it climbed (wicked) up the strip, so that the time for the water to climb 8 centimetres, marked on the strip, could be timed with a stopwatch.

I measured the wicking time for 20 pairs of "love" and "hate" water. Each pair of water samples was run at the same time to minimize atmospheric differences. After nine pairs had been run, new batches of "love" and "hate" water samples were treated by the same music as before and 11 pairs of wicking tests were run on the new water samples.

Individual results are shown in the table in Appendix 16. Summarizing the results:

- The "love" water took longer than the "hate" water to wick up 8 centimetres of filter paper in 16 out of 20 runs.
- For the first batch of music-treated water, the average difference in time to wick up 8 centimetres was 9.7% and for the second batch the difference was 6.1%.

The results show clearly that there is a consistent physical difference between the "love" and "hate"-treated water samples.

These results are opposite to those expected. As described earlier, in flow through cells, the smaller oligomers, created by love treatment, flow faster through the tiny cell orifices. But in the wicking experiments the love-treated water was slower to flow through the gaps between fibres. However, the gaps are thousands of times bigger than the cell orifices. This suggests a different mechanism is operating in the paper fibre gaps and the cell orifices. When I looked up the formula (See Appendix 16) for capillary rise – which is the scientific name for wicking of liquids by paper – the water viscosity (how "thick" the water is) did not influence the pressure driving the wicking. But the pressure was proportional to two times the surface tension (the tendency for liquids to act as if they have a "skin"), and it seems likely that the surface tension will be bigger if the water consists of big oligomers.

So I conclude from this brief experiment:

- There is a consistent physical difference between water treated with hard rock and gentle classical music.
- The surface tension of the water sample treated with hard rock music appears to be slightly higher than that of the classical music-treated water on account of the larger oligomers in the former and this resulted in faster wicking up the strips of filter paper.

Water as a Recording Medium

We are indebted to Dr. Emoto for pointing out the correlation between the quality of water and human (and animal and plant) health. This theme – the connection between water and health – is given an in-depth treatment by a number of specialists in the field in Dr. Emoto's book, *The Healing Power of Water* (Ref. 6-3). If you want to get into leading-edge science of water, I recommend it. From it I have learned:

- In material science structure is much more important than composition. Prof. Rustum Roy makes this point by referring to the element carbon. It is very soft as graphite (the "lead" in pencils), but if the carbon atoms are rearranged it becomes the hardest material we know on Earth – diamond.
- Prof. Roy points out that water can become a strong antibiotic if one part per million of metallic silver is dispersed in it – a fact attested to by sales from health food stores.
- Water exists as a multitude of structures which can be changed by purely physical effects – including human intent and music.

- In Dr. Emoto's book several authors (for example D. Knight and J. Stromberg) sum up the research by saying water's many structures give it the ability to store information.

These special properties of water make the preparation of homeopathic remedies possible as you can read in the next section of this book.

Homeopathy and the Crop Formation Analogy

The Angels continue their account of other ways energy can be stored in water (from Appendix 13):

> "Now we want to tell you about homeopathy preparations and this is independent of the discussion of oligomers, hexamers and the purity of water influencing the life of plants and animals. In addition to those effects, energy stamps can be put on water and these energy stamps come in a variety of different frequencies of love energy. Basically, they are a message: the energy stamp contains information that is different for different applications. The energy superimposed on the molecules puts them in a pattern. These patterns may consist of hundreds of thousands of molecules – it's a much larger-scale effect than the hexamers and the things we have just been discussing. The best analogy we can describe here is that of crop formations – the dimensions and the detailed structure of the crop formations are an expression of information in a message. Very often those messages are meaningful to humans who see the crop formations. They are formed by an energy stamp being placed on millions and millions of wheat stalks so that they become visible to humans and humans understand at some deep emotional level the message of the crop formations. The Emoto effect and everything we have been talking about up to now is on a relatively small scale of up to 1,000 molecules. Whereas, when we come back to homeopathy, the energy stamp

that is put into water to form the equivalent of a crop formation – a molecular formation – involves hundreds of thousands, and maybe millions, of water molecules."

"We can draw an analogy here between crop formations of millions of wheat stalks and the situation that occurs in a homeopathic preparation where an energy stamp has been put on maybe millions of water molecules so that they form very large groups or molecular formations. They are not constrained by hydrogen bonds, so it is not an enormous structure that has difficulty in moving through the water. It's just that the intent that is imposed by the energy stamp on the water molecules makes them want to be in a form that the energy has expressed by the structure. Individual molecules in the structure are quite free to move about and pass through cells in bodies of animals and plants, but in moving through the cells, they release that energy that is imposed on them. It is difficult to visualize the structure that is imposed by the energy stamp on the water molecules. It's a memory effect. The molecules of water in homeopathic preparations have a tendency to form into structures equivalent to crop formations. An energy stamp is quite fluid, it has intent to form, and reform, a particular structure, but the molecules do not have to stay in that structure as they work through a human body."

Water as a Recording Medium of Healing Love Energy

"We give the analogy of a tape recording of a piece of music. The energy of the music is imposed on a length of tape (a complete symphony can take up a complete tape) and the energy is dispersed among the iron oxide particles that are recording the magnetic signal. There is some order established in the magnetic memory of the individual iron oxide particles in the coating on the tape's surface. Viewed overall, if we could visualize the patterns that are imposed on the iron oxide particles, we would see a beautiful design that is equivalent to the molecular 'crop formation' in the homeopathic remedy. We have an analogy here, the crop formation is a static structure but when the crop is harvested that energy pattern is still present in the crop, in the wheat ears for example, and it goes into the preparation of bread, beer and food in general. Humans absorb the energy and it is beneficial to them; that energy comes into their food and bodies as love energy."

"Coming back to the tape recording analogy, we have the energy of a beautiful symphony that is imposed on the iron oxide particles of a tape recording. That tape recording appears almost random when it's viewed on a small scale, but if you could view the patterns of energy in the whole recording from one end of the tape to the other, you would see a beautiful pattern which represents the music that is recorded. In the same way, the energy stamp that is recorded on the molecules in the homeopathic remedy is spread out over millions of molecules. If you could view the pattern from a distance, you would see the beauty of the homeopathic remedy – the beauty of the fluctuations, the patterns of love energy that are recorded in the molecules. The analogy here is that the water molecules record a beautiful, beneficial pattern of love energy, which is directly analogous to the magnetic patterns imposed on iron oxide particles on the surface of a recording tape. Another way of putting this is to say that the water molecules in the homeopathic remedy carry with them the message of the remedy. For the remedy to have effect, it is necessary for the whole recording to be absorbed by the patient that is taking it. This is equivalent to saying the whole of the formation that is imposed on the molecules of water needs to be absorbed by the person to get the benefit of the energy pattern imposed by the energy stamp." (Also, that person needs a willingness to be healed, as in all healing modalities. – MKS)

Persons Interfering with the Homeopathic Process

"There is one other aspect that we didn't cover and that refers to an account in the book by McTaggert *(The Field)* (Ref. 6-4), which we referred to in our earlier communication. In the preparation of these energy stamps on water or some other molecules, if a person with very strong mental characteristics – that could be interpreted as negative or positive – is present during the imposition of the energy stamp into the molecules, then that person can interfere with the process. There was an account, in the book by McTaggert we have just referred to, of a woman who apparently spoiled the results of a research project on homeopathic preparations because she had a strong mental influence on them. It was realized that her influence negated the results of the tests that were being done in just that one case. When she was removed from the situation where she could influence the results and all the other people involved had no influence on the absorption of the energy stamp by the molecules, then the preparations were successful homeopathic remedies."

More on Solid Homeopathic Remedies and Mechanisms

"There was one little bit that we could tell you, Malcolm, were thinking about during the channeling and that was not all homeopathic remedies are liquids – made of solutions in water. Some are based on solid material which is compressed into a pill. You yourself took some based on sodium chloride (salt) as the recording material. The crystalline structure of sodium chloride or similar materials can be used to store energy – Big Surprise! This is what you dealt with at the beginning of this book, in which we talk about energy that is stored in the crystal lattice structure. Coming back to the sodium chloride-based homeopathic remedies, in that case the energy stamp that is put upon the sodium chloride is just like the one that is put on water molecules – it's just a different recording medium." (MKS: There's more on this in Chapter 7.) "There are several other chemical compounds that can act as recording media, but the principle is the same. An energy stamp is superimposed on molecules, which carry the message into the body of the patient who is taking the remedy. The energy stamp is transferred to the cells of the body that are doing the healing work. We don't think we can get into the actual mechanism of the energy stamp's beneficial effect on the structure of the human body – just let's say that it's like healing done by humans on each other. We will give more information on this in Chapter 7. The healing given by humans to each other is based on spiral love energy known as Qi (Chi) or Prana and described in our first book (Ref. 6-4). People that practice Qi Gong send energy patterns into a patient's body to heal them. In the case of homeopathic remedies, a similar healing energy pattern has been put into the carrier molecules that are absorbed by the patient. It's just like a Qi Gong person sending energy into a patient's body, except it's been put into a molecular structure, which is put into a bottle. That bottle is sold to the patient who takes the medication and receives the energy stamp that is healing to that person."

References for Chapter 6

Ref. 6-1   *The Hidden Messages in Water* by Masaru Emoto

Ref. 6-2   *Spiritual Chemistry* by Malcolm K. Smith

Ref. 6-3   *The Healing Power of Water* by Masaru Emoto (With contributions from experts in water science)

Ref. 6-4   *The Field* by Lynne McTaggart

Chapter 7: Healing with the Electromagnetic Song of Atoms and Molecules

In this chapter:

- ❖ *Coloured Light*
- ❖ *Spectra*
- ❖ *Interplanetary Analysis*
- ❖ *The Beneficial Song of Some Minerals*
- ❖ *Three Disease Treatment Methods*
- ❖ *Healing Jewelry*
- ❖ *Some Details of the Healing Mechanism*
- ❖ *A New Research Trend*

In the last chapter, the Angels and I talked about the storage of information energy in water. Now we are going to discuss energy stored in, and radiating from, solid materials. This radiated energy is what the Angels call the "song" of the atoms and molecules that make up the material.

(Chemistry Reminder: Atoms bond together to form molecules, e.g., one atom of oxygen bonds to two atoms of hydrogen to form a molecule of water.)

Coloured Light

When I watch a firework display I pay particular attention to the rockets that send a trail of red light in the sky. Because of my chemistry training, I know that atoms of strontium make the rockets' fire red. From my place on the ground, I can pick out which rockets contain strontium because they are signaling to me with their colour. That coloured light signal is part of what the Angels call the "song" of atoms and molecules, in this case, the song of strontium atoms.

Spectra

White light is made up of a mixture of different colours – the colours of the visible spectrum – which each have different wavelengths and frequencies: red (the lowest frequency and longest wavelength), orange, yellow, green, blue, indigo and violet (the highest frequency and shortest wavelength). But there are no boundaries between the colours, they blend together, like the colours of the rainbow, to form a continuous spectrum.

The radiation coming to Earth from the sun also consists of other electromagnetic radiation that we can't see, for example, the ultra-violet (which is at a higher frequency than violet), but we become aware of it when we get sunburned.

I will let the Angels explain (Appendix 14):

> "All materials give off electromagnetic radiation due to the vibration of electrons in the molecules. But humans can only see the radiation in the visible range – materials that radiate outside it appear invisible. Nevertheless they are giving off a signal and, if you have an instrument such as a spectrophotometer that can measure energy given off at all wavelengths of interest, then you can pick up their signal."

"An example we would give is in the ultraviolet range of the electromagnetic spectrum. Humans are not able to see anything visually, but you are able to measure it and so detect a signal from the chemical, a characteristic fingerprint of the chemical. That fingerprint is the 'song' of the electrons – or the characteristic tune the electrons play with electromagnetic vibration – that arises in the movement of the electrons that bond the atoms together in a particular kind of molecule. That 'song' is listened to by your instrument which can detect a particular chemical by comparing it with patterns of vibration made by known chemicals – songs that have been sung by known chemicals – with all the structural analysis that goes on behind the scenes in this science. You can then put together a spectrum of electromagnetic signals given off by the molecules and from the spectrum you can identify the chemical."

With the Angels' permission, I would like to break in here to give an example of how a spectrum can be used to identify a chemical. When I was doing research at an English university, I was studying crystallization of films of dye on small glass plates. In the case of one of the dyes called aminoanthraquinone, I could see flashes of light reflecting off crystals in the dye film. Microscopic examination showed the flashes were reflections off crystal whiskers – tiny ribbon-like crystals – standing perpendicular to the dye film which had already crystallized horizontally on the surface of the glass plate. There was a lot of interest at the time (1960) in whisker crystals in connection with electronic circuits so it was decided that I should analyze some of the whiskers. The story of how my guides led me to construct a miniature vacuum cleaner is told in our first book (Ref. 7-1). When I had constructed it out of glass tubing, I found that by working under a microscope, I could snap the whiskers off the dye film and the vacuum would suck them into a small tube of alcohol where they dissolved. I then ran a spectrophotometer test of the spectrum – the amount of light coming through the solution over a range of wavelengths

– of the unknown whisker material. Immediately when we saw the spectral curve, we recognized the whisker material as anthraquinone – a side product in the synthesis of aminoanthraquinone. The impurity could not find room to crystallize in the dye film so it "decided" to go vertical and form whisker crystals.

Interplanetary Analysis

The Angels continue to explain (from Appendix 14):

> "Scientists of the 18th and 19th centuries realized that all molecules give off a characteristic electromagnetic song. By tuning into that song they could obtain information on the composition, structure and what atoms are involved in materials that are lying on planets millions of miles away from them. That is a great benefit in understanding the origins of humans and the place they occupy in the universe. For example, in the case of the planet Mars, at the present time and over the last few decades there has been a lot of research looking for water. The research uses the characteristic electromagnetic song of water molecules to tell the scientists whether there is any on the surface of the planet. This was during the time that humans could not send machines to Mars – we know that they have now – but before that was possible, this was the only way they could determine whether there was any evidence of water on Mars."

> "This method of molecule exploration has been extended to other planets and star systems. The colour of the light – measured over a range of light wavelengths – that comes from other planets is the same as the thing we are calling the electromagnetic song. This electromagnetic song is broadcast throughout the universe by the molecules on another planet – let us say Neptune for example – and scientists on Earth can receive that light signal,

that electromagnetic song, and by analyzing it determine the composition of molecules on the surface of Neptune. This is an extremely useful tool in research and is one of the ways humans are starting to understand the structure of the universe in which they find themselves. It is a great benefit for humans to be able to measure these electromagnetic song signals, and it has helped make understandable the place of humans in the universe and how the universe around you relates to your home planet Earth."

The Beneficial Song of Some Minerals

"Over many centuries of experience, humans have come to realize that certain minerals have beneficial properties for their bodies, especially if the mineral is close to the body or even in some cases taken into the body in the form of powders. This is because the electromagnetic song is like the energy stamp that is put on water in homeopathy. In fact, in some cases that is the way the electromagnetic song is delivered to the body. By mixing a certain molecule with the water, the energy stamp from the healing material, which is put in as the 'seed' chemical, is recorded by the water's structure and is transmitted to the body of the person taking the remedy. That is what we talked about in Chapter 6. Now in Chapter 7, we are going to talk about how the material that is the originator of the energy stamp is able to directly influence the human body's health. The energy stamp is ever present in the molecule that is to treat the human health condition. If the treatment is not to be through water, then it may be through some other solid material, such as salt, or it may be conveyed directly from the originating molecule right into the human body. So here we have a gradation of treatment methods that we will enumerate." (Appendix 14)

Three Disease Treatment Methods

"First of all, we define our terms. The healing (or seed) molecule is the term we use for the material that brings the appropriate electromagnetic song to the molecules in the human body that require help to recover from the condition known as disease. The original healing molecule sends its electromagnetic song into the receiving molecules of the human body, and that song changes the structure of the receiving molecules and the cells that contain the receiving molecules; in effect, a healing takes place. The healing effect is complex, but we will give some more details in a later section. Generally, it is understood by humans that when healing has taken place, the structure changes in a way that we talked about in the first book (Ref. 7-1). So this conveying of the electromagnetic song is the all-important step, and it can be done in a number of different ways:

- It can be done by imprinting the energy stamp of the healing molecule onto water and that is what we talked about in Chapter 6.

- It can be transmitted through some other chemical substance, which is usually a solid, because there are not many liquids that are suitable for use as the transmitting material other than water, of course, which is a prime example, with its strong ability to record energy stamps. Mainly in this second category, the transfer of electromagnetic song is through a solid carrier material such as sodium chloride, in which the song of the healing molecule is recorded. The soluble sodium chloride is taken by mouth into the body of the person to be healed and the energy stamp is delivered to the appropriate parts.

- The third method is direct application of the healing molecule to the human body. On rare occasions, this is done by grinding up the healing material and putting it in tablet form, which is then put in the human body where it goes to the required location, and delivers its electromagnetic song of healing. But more often, the electromagnetic song can be broadcast from the surface of the body into the area where the treatment is required. This is the case of the healing molecule being in a material that you would call a mineral or maybe a semi-precious stone. Through many years of practical experimentation, the electromagnetic song of any particular mineral has become recognized as being efficient in healing a particular disease. That knowledge is contained in many books such as the one you have Malcolm (Ref. 7-2), on the use of minerals to heal diseased conditions of human or animal bodies. You have used that book to find the appropriate stone, which was black tourmaline, that you made into a necklace for Brenda to wear and help heal her cancer. In those cases, the stone is put in close proximity to the body that requires the treatment, and the electromagnetic song goes directly from the healing molecules straight into the body – we would say the song is broadcast into the body – and the body responds by receiving the signal and being healed." (Appendix 14)

(For success in all healing modalities the person to be healed needs to give intent to be healed – MKS)

Healing Jewelry

"It is a common human practice to take those healing minerals, confer on them titles like 'Healer of Cancer' and fashion them into jewelry that can be worn on the body. This is a convenient way of broadcasting the message from the healing molecule into the body – and it works very well, as you know from Brenda's experience. It was not accidental that Brenda's need came up at that particular time. It was planned that there would be a demonstration of the efficiency of jewelry of appropriate mineral content in healing. This was set up for you before you knew about this particular book, but now looking back you can see the point of having this demonstration. So that is how minerals and semi-precious stones are able to heal human diseases. It's like homeopathy, except there's no intermediary material to record the energy stamp. In this case the energy stamp is delivered directly from the healing material directly into the body by a broadcast of the electromagnetic song. You could say it's like having the singer of the song broadcasting the healing song directly into the human body rather than a recording being made in water, salt or some other material and then playing that recording inside the human body. So there's an explanation for the 1000-year-old folklore which tells which minerals are able to heal certain human diseases. That's all set out in encyclopedia of healing stones like the one you possess, Malcolm." (Ref. 7-2)

"We think that's rather a good analogy: molecules doing the healing are singing their electromagnetic song all the time. We can record that song in water or salt and deliver the recording into the body or apply it to the body in some way. For example, the healing molecule can be worn close to the body in the form of jewelry, carried in a pocket or placed on the body during meditation. The person who wants to be healed can receive the benefit of the electromagnetic song of the healing molecules directly into their bodies. It's a very efficient method that works well." (Appendix 14)

Some Details of the Healing Mechanism

"Perhaps we should say a little more about the actual process of the healing. The energy or electromagnetic song that's given off by the healing molecule is received by the disease-centered molecules. The electromagnetic song consists of layers of frequencies. The frequencies that you humans measure as the electromagnetic song, the electromagnetic radiation in your normal range, are accompanied by overtones, the frequencies of which occur in a fractal pattern. They are fractal frequency overtones that range from normal frequencies of electromagnetic radiation, that originated in the movement of the molecules, right up to the frequency of spiral love energy, which you know is 10^{33} hertz. When an electromagnetic song is broadcast via a mineral into a human body, the electromagnetic part of the broadcast is accompanied by fractal overtones, which do the actual healing."

"It's the same sort of energy that occurs in the energy field you develop when you combine two vibrations, like in your walking meditations when you combine *yin* earth energy and *yang* sky energy to create a scalar energy field (Ref. 7-3), which has frequencies that include spiral love energy. Those same frequencies corresponding to spiral love energy are present in the signals of the healing molecules and in the minerals that are used as healing jewelry. The same energy of very high frequency is present in the energy stamp that is conveyed by the water or solid homeopathic recording material into the body. Accompanying those frequencies that humans regularly measure are other frequencies corresponding to scalar energy fields and spiral love energy at frequencies of the order of 10^{33} hertz. It is those overtones which occur in the whole spectrum of frequencies emitted by the healing molecules which relate to each other in a fractal way. It is the higher frequency component of the spectrum that is the actual healing energy. It is the same kind of energy that is summoned by a Qigong master and directed into the body for healing purposes." (Appendix 14)

A New Research Trend

I will let the Angels finish this chapter with hope for a new kind of research (from Appendix 14):

> "Most regular doctors tend to laugh about this sort of thing, so do conventional scientists, because they can't see how the electromagnetic song can possibly heal a human body, because they are so fixated on the idea of having to put chemicals into it. Some people in responsible positions are beginning to understand there are more ways to deliver healing energy stamps than have been realized in the past. In what has been regarded as superstitious knowledge and associated with witchcraft and psychic charlatans, there is in fact a basis of science. It is a new science that is not generally recognized, but some people responsible for health issues like this in the human situation are beginning to realize that there is more to the folklore on minerals for healing than has been acknowledged in the past."

The book, *Tuning the Diamonds* (Ref. 7-3), mentions some people involved in this new kind of research. I will list some of them here so that interested readers can follow their work:

- Victor Stepanovitch Grebennikov – Discoverer of the "Cavernous Structures Effect."

- Dr. Valerie Hunt – Reports on "clinical evidence for bio-scalar healing."

- Dr. Barbara Ivanova – "Body, mind and soul must be brought into the process" (of healing).

- Glen Rein, Ph.D. – A leading pioneer of bioelectromagnetics concluded that an electromagnetic field by itself will do some healing, but when you add a scalar component the biological system really takes off!

References for Chapter 7

Ref. 7-1   *Spiritual Chemistry* by Malcolm K. Smith

Ref. 7-2   *Healing Crystals and Gemstones* by Flora Peschek-Böhmer & Gisela Schreiber

Ref. 7-3   *Tuning the Diamonds* by Susan Joy Rennison

Epilogue

The Angels of Crystal Light and I wish to add the following from Appendix 12:

> "Now there are the first instances of scientists realizing that it's important to take into account God, His creation of the universe and the love that flows throughout that wonderful universe. Without understanding of that love, scientists cannot pursue their investigations that are concerned with only physical things. It is not possible to explain the universe unless you take into account God and the love He has given the universe. That love is the essential driving force of the whole universe as humans perceive it. Without love it is not explainable. But you, Malcolm, have made a good start in our first book in putting that value of love forward as an important scientific 'variable,' if you like. You said in the introduction to our first book that 'Science doesn't make sense without God.' We hope that as a result of the books that we are going to write, that lesson will come through and scientists, and all people on Earth, will understand that the universe and its structure, its meaning and its purpose, are only explainable in terms of God, All That Is and the wonderful love He showers upon us all."

The following pages give the appendices
I have been quoting throughout the book.

Appendix 16 records results of my experiments to detect a
physical difference between "love" and "hate" water.
All the rest are the communications channeled from the
group of Angels called Crystal Light.

Appendix 1:
Angel Communication, November 8, 2009

Hello dear Malcolm, we all love you for the work you are doing. Completing the book (*Spiritual Chemistry*) and getting it ready for publication was a big step, and we are very happy with the way it went and the result which you are going to give out to the world very soon. Because we know that you enjoy doing this work so much, we thought rather than you feeling at a loss because you have no writing to do, we would come and give you some information which you can use in your next book which, as you have rightly guessed, is about energy being stored in objects and materials.

You got that preview this morning, just after breakfast, about energy being stored in relics such as the blood and bones of ancient saints, people who lived very holy lives and who were revered for their holiness. The energy that was brought in by the spirit residing in the body of the saint was so strong, it affected the energy and molecular structure of the body cells and gave them remarkable properties. Some of these remarkable properties were seen after the death of the people concerned.

For example, Saint Theresa of Avila, who you know is generally called the Flying Nun because of her ability to overcome gravity – but that is a topic we will talk to you about on another occasion – coming back to Saint Theresa, when she died, her body was not consumed by the usual decay that occurs with human bodies. Her body remained in a very pure, unaffected state for hundreds of years after her death. It has now rotted away since 100 years or more but for the first 200 or 300 years after her death, her body remained as it was after her death. This was because the great energy stored in the cells of the body preserved it. This is something that medieval people recognized. They knew that the blood and bones of the saints were special because of the energy that resided within them. They made great efforts to bring those relics into special churches and cathedrals.

You, Malcolm, have been on tours of Spain where in certain churches you have seen vials of blood. The blood in the vials is reputed to liquefy at certain times during great religious ceremonies – this is true. We know you haven't seen it with your eyes but there are many church people who have seen the blood liquefy, and they tell all the people that are gathered in the church what has happened. The gathered people have to take the word of the priests because they can't see it for themselves.

This happens because the small samples of blood of certain saints are infiltrated with the great energy that the spirits of the saints put into their bodies. This energy gave the blood apparently miraculous powers, such as liquefying at certain times. In a similar way, bones of saints are full of the energy of the saints, and are very special in that respect. Devout people are aware of that energy and it draws them to the relics. So it's not just a question of an old superstition by medieval people thinking this was like some magic conferred on the saints' bones. Although if we look at it in the light of 21st century knowledge, it does seem like magic that dried blood will turn into a liquid form and back again into a solid on particular occasions. These are all phenomena that are resulting from special love energy that the saints' spirits conferred on the material of the bodies they occupied.

In a similar way – not part of the body exactly but very close to it – were certain woven materials like Veronica's veil, the Shroud of Turin and other similar pieces of textile material that were in close proximity to the bodies of saints or, in the latter case, in close proximity to the body of Jesus. Those textiles were influenced by the great angelic love energy that the wearer conferred upon them. In some cases, they have apparently miraculous prints of facial features and/or wounds on the bodies that were conferred on them by the close proximity of the textile to the saint, whose energy brought about chemical changes in the textile. These are very similar to the miraculous prints that we talked about in our first book. Of course, Veronica's veil – although it was named after Veronica – really has the print of Jesus' face on it, because that was the person who had that great love energy which affected the textile and conferred on it the design that was recognized as the face of Jesus. All these examples we are quoting now, and there are many more, are ones that are known to you Malcolm. You can do some research on this type of material and find lots of other instances of this kind of material. A good place to start is the book *Miracles* by D. Scott Rogo.

If you were able to approach close enough to these materials – but it is unlikely that the church would allow you to do so – you would be able to detect with your pendulum the very high energy that is stored in these materials. You would realize that they are not just attractions that brought people who seemed to be like tourists in their time but who really were pilgrims – at least that's how they thought of themselves – to places like Santiago de Compostela. There, they were able to come close to the miraculous high energy of the saints. They would travel many miles on foot to be near the relics of the saints. To them that was not so much a physical holiday, because it was very much like their everyday life, but it was a spiritual holiday. They gained a great deal of energy from the pilgrimage that they made to be in close proximity to the relics of the saints.

You can do a lot of research about this, you don't need to be really close to these objects. You saw in the book by Laszlo that dowsers were able to pick up on thought patterns set into the ether (we'll call it). Those thought patterns were set up in the four dimensional space-time that you occupy by tides of energy in the 5th dimension, so the actual original energy that is causing these reactions is in the 5th dimension. People with pendulums and other dowsing instruments can detect those energy sources remotely. For example, if there is a relic at Santiago de

Compostela, it is not necessary for you to go there again – we know you have been there – in order to dowse the energy that is held there. Instead you can do it sitting in the same chair that you are in talking to us now. You can focus your intent on Santiago de Compostela – and this requires some concentrated studying of the history of the relics and a description of the relics, we know you are able to do that kind of research. When you have established a connection with that place, which will be through the cosmic lattice, then you can dowse and do experiments, just as if you were dowsing the energy of a picture that is before you. We know you don't know how to do that yet, but that is something that will be coming in later talks that we give.

You have probably guessed by now that your next book will be about the energy that can be stored in objects and materials. We will show you many different examples of this energy such as: Dr. Emoto's water samples – the water samples that you created and measured Malcolm – and energy in pictures which you and Michelle have talked about. All these things you can research and you can approach in their template in the 5th dimension. You should know the meaning of that now from the writing in our first book. You can measure that energy by devices that we will show you. As you know, one of the most likely devices, besides your pendulum, is the torsional pendulum. We know you are familiar with how to make those and you have an article about measuring weak forces such as Tesla fields, or scalar fields – the energy fields you get by combining two types of energy. Those fields are created every time you do a walking meditation. The other form of them – but with spin added – is the torsion field, which has become a Russian specialization. (A good start in studying these fields can be made with the article in the Nexus magazine.)

All those fields, and the energies that create those fields around them, can be measured by pendulums, particularly the torsional pendulum. The advantage of the torsional pendulum is that the strength of the energy field, which is applied to the pendulum in the form of a spiral, relates directly to the deviation of the pendulum from the zero position which can be measured as the strength of the field. We suggest that you build a torsional pendulum that you can move around from place to place, so it would need to be in a small box with some transparent sides to measure the deviation of the pendulum. Place this in a known geographical location with its zero position in a certain compass bearing, then by measuring the deviation of the pendulum from the zero position, you can measure the strength of the torsional field that is exerted by the special materials.

Now you have an introduction to our second book, you have a general idea of what it is about, and you have an introductory chapter to it in terms of the miracle materials and relics of the saints and how those objects retain special energy, which can be measured. With that brief communication we will close now. We just wanted you, Malcolm, to have something to work on and to think about as you go about the publication of our first book and the commercial processes that you must go through in order to get it into the hands of many people.

Thank you for doing this for us again. We know you enjoy this work and you like to be part of some big study all the time, so we know that you will enjoy doing this. We thank you for all your efforts on our behalf. You will have another adventure into book writing in the next year, and it will be just as enjoyable as was the creation of the past one. We leave you now and love you now and wish you a very happy day. With all the blessings that are ours to give we leave you now,

Crystal Light.

Malcolm K. Smith / May 1, 2010

Appendix 2:
Angel Communication, June 18, 2010

Hello dear Malcolm, hello! Wonderful to be back with you again doing this! We see a big smile on your face, we know that you love doing this work. We are very happy to be able to come back to you for a second book. Your first was very successful, and is being very successful still. It is making inroads all over the world; you are not aware of the success because you can't see it. Very soon, you will see reports from many different people how successful it is. We are helping the book move forward from our side as well, making sure that it reaches many people who can influence other people to read it and learn about the state of the universe and how it all works, which you have laid out so carefully.

We are very happy with the progress that is being made, that is why we now come together to start the preparations for another book, which as your Guides have told you today is to be called "Hunt for Energy in Objects", although that is a working title, and we don't know if it will be the one that will hit the book stores eventually. We already gave you some information on this topic in connection with the energy of saints' bones and relics. We did that in November when you were feeling that you missed channeling with us and you were struggling to get our first book finished. At that time, we knew you needed a boost of energy, so we gave you the talk about energy in saints' relics to help you through that phase of getting your number one book on the book shelves. It worked and you were refreshed and energized by that talk we gave, so that will be an important part of the work that will come together with what we are giving you today and on subsequent occasions.

Today we want to talk about another topic which you realized was coming up, and that is the energy in crystals. Because this is one of the sources of energy which is probably the most easily observed, measured and recorded of all the sources of energy in objects. This is because crystals have a unique structure. As you know as a chemist, they are very ordered structures with the atoms being arranged in a very regimented way like soldiers on parade – or Marine Cadets on parade if that analogy appeals to you more – and all those atoms stand at the same distance from each other and form a very regular matrix. It is into that matrix that energy from other sources can be inserted and stored. The energy of the atoms, when the crystal is at normal temperatures, is very stable – the atoms, as you know, vibrate a little about their mean position. In that way, the atoms form a sort of frame which allows other energy to come in and be inserted in the frame, so the energy that comes in is held and vibrates within the framework of the atoms of the crystal.

My Guides ask me to insert at this point:

The atoms are stable at zero point energy.
Residual energy makes them move slightly.
The atoms do not move a lot until the crystal melts.

An analogy that comes to our mind that would make sense to you on Earth is that these crystal atoms are like picture frames. The energy of the atoms of the crystal form the frame itself – this is a three-dimensional frame, not two-dimensional as you would have on a wall in your world – and this three-dimensional frame is quite stable and strong and is able to hold a lot of information. That information comes in from outside sources, say from some event that occurs around the crystal, and that energy is then held like the picture in the picture frame. It is held between the energies of the atoms around it, which forms a very stable cell.

Another analogy you could think of is that each group of adjacent atoms in the crystal – which is referred to as a unit cell in crystallographic terms – is like a living biological cell into which information can be inserted, just as a paramecium organism brings food into its single cell body and holds it and uses it. In the same way, you can regard the crystal as a living organism that "digests" information, although it doesn't break it down, it holds it until it is ready to be released for some other application.

As you know, there are people on Earth who are able to relate to the energy of crystals. They are able to pick up energy from various sources and store it in crystals, and then to read it while it is still in the crystal. We refer you to some of the books written by human experts in that area. We would like you to read those books and become more familiar with the things that people say they can do. (As I write this part I "see" in my mind's eye the book, *Psychic Discoveries Behind the Iron Curtain* – MKS.) You will find that our explanation goes along with the facts that those people present in terms of what it is that a crystal can do: how a crystal can absorb energy, how it can store it and how it can give it up. Although it is not so much that the energy is given up, but once it is in the crystal how it can be read by people who can tune their senses to that particular frequency of the energy that is stored in the crystal frame.

You have used crystals yourself and felt that there was some special energy there. In particular, you had your crystal of calcite on a pendulum with which you channeled our first messages in 2003. You did not know that we were the group called Crystal Light at that time, you thought it was your Guides who were speaking to you through the crystal. But in fact, it was the group that is now addressing you and we are known as Crystal Light. We are able to come into the framework of your crystal pendulum, which you still have, and you can bring that out and start using it again if you wish. (I was shown it by my Guides right after this channeling and I started to use it again – MKS.) We have good communications with you, so you do not need to do that but if you would like to for sake of tradition, then you could do that.

Coming back to our relation to that crystal; we Angels are able to move into the structure of the crystal – as you know the cosmic lattice is everywhere and it feeds into the gaps between the crystal atoms, into the frames formed by the crystal atoms. We take that route following through with the cosmic lattice into the crystal, and are able to present information. We can influence the swinging of the crystal by the connection that it makes with your muscles and the DNA in your body cells. You have already written about this, and are quite familiar with the mechanism of the communication between dimensions. The crystal is an added tool for making that connection more easily manifested. By coming into the crystal and being within its energy framework, we are able to influence how you swing the crystal and so that is translated into movement of the crystal around your alphanumeric chart, very efficiently recording words and concepts that come from us originally. The poem "44" was written in that very way and in a short time considering the length of the poem. All the words in that poem came through the cosmic lattice, though the crystal into your body, and your body interacted with the swinging of the crystal to spell out the words in a very fast, efficient way.

Sometimes it is necessary to remove information from crystals, they become saturated with energy at times. That is best done by intent of the person who owns the crystal and uses it. You have heard about people running water over the crystals to do that, but the water is more symbolic than actually washing out of energy. The energy in the crystal is not impacted by the water, but the intent by the owner of the crystal to flush it out, this is facilitated by the ceremony of washing the crystal or leaving it to stand in water. (As I write this I see parallels to ceremony in healing – MKS.) Of course, many people do not realize that some crystals are soluble in water – they may have grown from an aqueous medium – if you put too much water on the crystal, you may damage its surface. We don't need to tell you about the chemistry of those materials that make up the crystal. You know about that from your chemistry studies as a young man.

(The release of energy that we were just talking about when

people wash their crystals is release driven by human intent. It is different to the lasing action of crystals when the unit cells are so full of energy that they spontaneously release it.)

The inputting of energy into crystals and the removal of energy from crystals is all part of an intent process – human intent or Angelic intent or intent of your guides – it's what puts energy into the crystal and can "read" it out and can remove it. There are three operations involving crystals and energy: you can put energy into the crystal or read energy in the crystal or remove energy from the crystal. Those are three separate operations and they are controlled by human intent or Angelic intent.

The cosmic lattice that you have discussed at length in our book is a pathway for many people to use in gaining information. All the billions of people that you have in your world are able to access the cosmic lattice, if they so wish, and if they know about it – and more and more are getting to know about it. If it was necessary the cosmic lattice could be used as a conduit to take energy into any particular crystal. That is what happened in the days of the first Atlantis. As you have told people in your talks, the Atlanteans had great abilities in growing very large crystals and they used them as power sources and information storage devices.

In those days, the cosmic lattice was as effective as it is today. Those Atlanteans that wanted to have a crystal with particular properties would come together mentally and give intent to put that energy into the crystal. They would be able to feed the energy through the cosmic lattice into the crystal and leave a very strong recording of the energy in the crystal. The energy/information could be read out of the crystal later as necessary. That energy could be used to control a further flow of energy that was beamed through the cosmic lattice into the crystal and then out from the crystal to space ships traveling high above the Earth. That energy was beamed through the crystal to the space ship as power to drive it.

Crystals can be used to generate energy beams. That is what was used as a source of power for space ships in Atlantean times. It is an analogous process that you have in your present time in which lasers are created from crystals. The original crystals that were used in your present Earth time were rubies, and you can look up the chemical structure of rubies and add in that piece of information. It is not necessary to have any particular crystal structure, it's just that some structures are better than others for the different processes we are talking about.

(MKS note: Earth lasers, such as those based on ruby crystals with half silvered mirrors at each end, work by atoms being raised to an excited state by "light pumping". When all the atoms are excited, they will suddenly release all the energy when a little more light energy is pumped in. We call it Light Amplification by Stimulated Emission of Radiation – LASER. The release of energy from over-full crystal unit cells is a slightly different process, as described in the following pendulum notes:

ADDED July 19, 2010:

We would like to talk about the alternative laser mechanism.
As you intuited earlier, there is an alternative mechanism for lasers produced by some crystals.
This involves all the unit cells in a crystal filling with energy first.
When full, the crystal can empty spontaneously as a little more energy is added.
The action is like a siphon.
We can only say that the energy is siphoned out of each cell by the next one emptying.
Energy is like a viscoelastic liquid it pulls energy out of adjacent cells.
This applies only to spiral love energy.
The elasticity between spiral energy coils is the result of zero point energy (ZPE) attraction.
Remember ZPE biases counter-clockwise spirals. [See Appendix 1 in Spiritual Chemistry.]

Energy coils stick together like parallel plates in the Casimir Effect.)

As you know from your chemistry background, different crystal structures are possible, for example sodium chloride produces a crystal structure with a cubic unit cell or frame. That is a structure which is particularly suitable for some kinds of energy. Another type of crystal structure is prismatic and that is suitable for a different kind of energy. Just as people of different races have bodies with different colours of the skin and different structures that are suited to the environment in which they live, so crystals have different structures which are suited to different applications of energy storage, reading and release.

We suggest Malcolm, that you construct a torsional pendulum. Then with that you could measure the energy released by crystals, or read the energy stored in crystals, by modification of the periodicity of the torsional pendulum. Say, for example, you have a torsional pendulum that is oscillating at 50 cycles per minute – you should also measure the amplitude of the oscillation. Then if you bring near that a crystal which has got information stored in it, you may find that the oscillations change to 60 or even 70 per minute. The amplitude of the oscillation may also change considerably in the presence of the crystal. That is the influence of the energy that is in the crystal interacting with the periodicity of the torsional pendulum. This will be a handy tool that you can develop and we think you already have a rough idea of how that should be set up.

People very often carry crystals around with them as indicators of energy situations. Just as we told you that the crystal pendulum you had responded to energy input through the cosmic lattice, so people can use crystals in combination with the cosmic lattice to detect certain energy situations. One of those is the use of a divining rod. When Michael gave you some prospector's equipment, included in that was a divining rod. Crystals of different kinds for different applications can be put in the divining rod. The purpose was that the divining rod was carried by the prospector with a crystal inserted in the cavity at the end. That divining rod would then dip when it received emanations from metallic sources in the ground – so it was a means of detecting metal ores in the ground. That is another area of study for you. Those divining rods have to be held in a particular way – very lightly and pivoted at a certain point – so that a small movement of the crystal can be amplified by a movement of the whole rod when a metal ore is detected. The prospector can detect that movement of the rod and realize there is an ore source nearby. That is a more practical application of crystals, but we use this as another example of the utility of crystals and the many different ways in which they can be applied.

Now we want to talk about the interaction of biological systems with crystals. You tell the children that birds, cetaceans and butterflies are able to make use of the Earth's magnetic field. That is all done through crystals in their bodies as well. In those cases, the crystals are of magnetite or some other similar chemical. Those crystals respond to a magnetic field and show the creature – the bird or cetacean – the direction of the north magnetic pole of the Earth, and from that they can deduce the route they need to follow for their migration. The interaction of the magnetic field with the magnetite crystal is directly analogous to the interaction of the cosmic lattice with a crystal in a pendulum, divining rod or just simply held by a person who is sensitive to the movement of a crystal.

Just as the birds and cetaceans are guided by the very small movements of a crystal under the influence of a magnetic field, so humans can detect small movements of their crystals in various devices to follow fields that are set up by the cosmic lattice. In this sense, the crystals are acting as amplifiers of fields that are not seen by the humans but can be detected by them. When Rupert Sheldrake talks about morphic fields, he is referring to the fields that you have suggested are scalar energy fields or torsion energy fields. Morphic fields consist of a scalar energy field which can have a rotation component at times, then it's called a torsion field, although we just consider them as a field. Crystals in biological systems respond to those fields – it's the same process that the birds and cetaceans are using when they respond to the magnetic fields around them.

All human body systems are based on crystals – for example cell walls are defined as lipid liquid crystals and the heredity material DNA has a crystalline structure that humans call the double helix – which are able to respond to fields around them. So we are talking about the same kind of process in biological systems as we were when talking about people using crystals in pendulums and divining rods. The crystals that are in the cells of your body, in particular the DNA crystals, are able to respond to the morphic fields that are around them. In responding, they can change their structure and this is the basis of healing that takes place in morphic fields or fields that have been put upon a human by someone that is healing. We mention this because this is a connecting point to the mechanism of healing, which we will talk about on a different occasion.

(MKS NOTE: When I was channeling the last two paragraphs my connection with the Angels faltered and as a result the concepts are not stated clearly. I asked for a clearer explanation through the pendulum and got the following:

There are three parts to this comparison.
First is the response of magnetite crystals to magnetic fields,
This occurs in birds, cetaceans and humans.
Second, crystals in devices like divining rods react to scalar and torsion fields.
Third, crystals in human cells like lipids and DNA react to scalar and torsion fields.
Crystals in all three situations are responding to fields.
Scalar and torsion fields are the basis of Sheldrake's morphic fields.

I then added a fourth similar situation. Referring to Figure 5-2 in *Spiritual Chemistry*, the microphotograph from my Ph.D. thesis, it can be seen that the dye molecules crystallized in a spiral pattern in response to the torsion field around the large crystal that was deposited on the glass plate during the early stages of the vacuum deposition. This can be interpreted as the crystals "moving" as they grew, i.e. the developing crystals were curved.

Coming back now to the healing course mentioned above:)

We mention this because this is a connecting point to the mechanism of healing, which we will talk about on a different occasion after you have taken your healing course. It's not by accident that you heard about that course, we intended you to go and take that course and learn more about healing. Then, from the information you gained from your course, you will have a better understanding of the processes by which energy is brought into the human body by a healer and can be used for self healing or healing another person. It all works through the crystal structure of the DNA and other similar macromolecules in the human body. They are able to bring in energy from fields and use it for changing the structure of the body, so that healing can be manifested as a result of the structural changes in that part of the human's body. This is a segue into the area of healing, and as you can see, there is a strong connection with crystals being used for healing and for storing energy.

When you were a boy, you were very interested in crystals and you used to make beautiful blue crystals from copper sulfate. This was because you were fascinated by the chemistry, but also you had a remembrance from previous lifetimes when you used crystals for information storage and for healing. One area that you could have followed in your research at Manchester University was the energy in the crystals that formed in your dye films. You have that beautiful example in our book of the photograph from your Ph.D. thesis of the crystals that grew in the dye film that registered the spiral energy field around that relatively big crystal of dye. That process goes on through many situations that you have explored.

You are beginning to detect the source of those "bush" crystals that you saw growing from the dye films. As you are beginning to realize, those are the subject of research on that other planet in the Sirius system which you call Kuzalini. You have seen in your interplanetary travels, people working on structures which you recognize as being very similar to the bush crystals. This is something that you will be able to focus on and will be the subject of another session that we will give you on energy and crystals. So you see, there are several branches of this study that we want to talk about. We are just giving you a brief overview of the different branches at this point but focusing, on this occasion, on the concept of the crystal unit cell, or frame, holding information within it – information that can be read out or taken out of the crystal just as easily as it is put into the crystal. Of course, you realize this is the basis of much work on computers, where gates are formed in crystal structures that are able to control the flow of particles such as electrons. Pieces of information can be stored in the crystal, can be moved around and read via the structures. This is all the basis of the big industry that you have in your world at this time of crystal structures used in electronic devices.

There are many applications for crystals in your present world: energy storage, information storage, healing and prospecting, to name but a few. There is much work to be done in sorting these out and coming up with a theory which is comprehensive and covers all these different aspects. We should also add into that the natural sensing of magnetic fields by birds and cetaceans, as well as another application of special crystals.

So Malcolm, we think we have given you enough for today. This is quite a broad brush-stroke description of the situation around crystals. It is a big area and you will need to study books and understand how all these things work from a mechanistic point of view. We know you can do that, because you are used to doing research. We will come to you again and talk about each of these applications in a more focused way so that you can flesh out the information that you gain from books. You will be able to build on a skeleton of book information a whole body of information that we provide to you as flesh on the bones of your original structure.

Thank you for doing this; we were not disturbed by the gardeners and their machines. We were glad you were able to find another location where the disturbance was less noticeable. Than you for doing this Malcolm, you are a very faithful worker for the light. We appreciate and love you for all the effort you put into this. We know we give you pleasure in talking about and showing you these things and bringing them to your attention. We are joyful that you are so happy to do this, because it matches the joy that we have in presenting it to you and to your world. We are very grateful to you for relaying these thoughts and putting them into words so that people in your world can read about all these wonderful gifts that come from All That Is.

Now we say goodbye; you can go and watch your football match and enjoy yourself knowing that the second book "Hunt for Energy in Objects", is already started. Thank you, Malcolm. We love you. We send you all the blessings that are ours to give. We will come back again soon. Goodbye.

Elapsed time 45 minutes.

Malcolm K. Smith / July 18, 2010

Appendix 3:
Angel Communication, July 24, 2010

Hello again Malcolm! Nice to have you here with us in this cool place where we can relax, talk and pass our messages to you. It's good to not be concerned about surroundings with no noise from gardeners this week. On this day of the week, everything is peaceful and calm, so we are ready to talk.

Today we want to come back to the topic of bones: bones of the saints and bones in human bodies as well. This is one of the more crystalline structures in the human body and therefore one of the places where vibrations, fields and influences of material energy are stored. Bones are made of a material called apatite – that's something you need to check (calcium phosphate $Ca_5(PO_4)_3X$ where X is fluorine, chlorine or OH ion – MKS). Of course, as you know from your dairy produce advertisements, calcium is the main element that occurs in bones. Crystals of this material reside in the bones and form their main structure. Of all the body structures, they are the strongest and most resistant to crushing and tearing. The skeleton serves a double purpose of maintaining the body, which you use as a vehicle, in its full position so that it can move around and function properly. In bones there are several different compounds, but the one we want to focus on today is apatite which forms crystals right throughout the human skeleton.

In addition to their strength, apatite crystals have a second purpose of storing within them energy. This is the premium storage place for energy which is created and received by the human spirit during a lifetime. As you know from the first book which we wrote together, most of the energy comes into the body via the DNA. The human spirit itself, which resides in the body contributes much of the energy that goes into the body.

Some of the energy goes to create the body, maintain its structure and grow new cells – the turnover of cells is necessary for the good health of the body.

There is a certain component of the energy that comes in which is, if you like, a recording of the experience of the body, and that recording goes into the crystals, which occur in the bones of the skeleton. That is the principal storage place for the energy of the experience of the human during a lifetime. That is why the bones of saints are so revered more than any other part, because the bones contain the most vibrations, the most energy that has come from the lives of the saints. Some people in present times and many people in past years were able to detect the presence of energy stored in the saints' bones. That is why the collection of saints' relics – they were called relics but basically it was the bones – were such an important part of religious practices particularly in medieval times.

Yesterday you Malcolm had a wish to go back to Santiago de Compostela in Northwestern Spain – and you will go there again as part of your research – as you know the bones of Saint James (Santiago in old Spanish) are stored in the cathedral there. That is why that place is so revered and why you are attracted to it as well Malcolm, because you were involved in the collection of the bones and putting them in that cathedral. As you know the legend has it that there were shepherds in a field and a star came down into the field, hence the name Compostela – field of the star. Of course in modern times, you realize that star was a UFO, or an alien space ship if you prefer a more definitive term. The aliens were cooperating with Angels like us, the Angels that are speaking through you now, and we showed those shepherds where Saint James's bones had been hidden.

The reason for the hiding of the bones was that they were on their way from the Holy Land to Europe, and pirates attacked the ship that was carrying them across the Mediterranean. The crew of the ship realized that the pirates were coming, and so a small detachment left with the bones in a fast small boat that made it to shore before the pirates got to the ship. That small detachment of the crew hid the bones in a safe place. The Angels and the aliens knew of the hiding place, and when all was calm and worldly threats were gone, the Angels arranged this meeting and reparation to the shepherds in the field of the star.

The craft was piloted by "Nordic" aliens and was just like the ones that you and Marjorie traveled in. That alien space craft came down in the field and the Angels told the shepherds about the location of the bones. The shepherds went to the hiding place, recovered the bones and took them to the place that is now called Santiago de Compostela. Of course, the shepherds were guided by the Angels to do this. In those days it was regarded as a miraculous event – it would still be regarded as a miraculous event in modern times. In some ways it resembled the miracle of Fatima.

You were steeped in those occasions, those meetings with aliens. You had experiences about Fatima which we won't go into now, but you know what we mean about the photo of the figure standing over the tree. The reason you are so involved is that you were one of the shepherds that was involved in the recovery of Saint James's bones. One of your "missions" in different lifetimes is to recover relics and to care for them in other lifetimes. Now in this lifetime your scientific side is coming to the fore and you are explaining why they are so important, the understanding and knowledge behind these events. This is something we want you to write about in your next book, the background behind the energy. Now, we go into how it is stored and the reasons for it.

Recalling out last session with you, we talked about crystals consisting of rows of atoms lined up like soldiers on parade. In between these atoms, which act like frames, information can be stored. We like the analogy you came up with – or maybe it was given to you – of music resembling the atoms in the crystal structure. The atoms represent the staff and bar lines, creating a frame, in which energy can be stored or held like a picture in a frame. In this analogy, the energy is represented by the music notes that are written on the lines of the staff. That is a very good analogy, because it has the feeling of energy coming from the energy of the music. We recommend that you use this analogy in your writing as well because it appeals to many people, particularly the non-scientific ones, who may not be sure of the structure of crystals, but they can see how the bar lines divide the music up into measures. We would say in this analogy each measure of the music – or bar as you would call it in England – represents one unit cell of the crystal. Around that unit cell are the atoms and the energy of the atoms creates the cage in which other energy can be placed, can be read out or can be removed. We covered all that in the last session.

So in bones, human bones, there are many, many, many of these musical measures that can store a lot of musical notes, representing the energy that has come to the human by the way we just described, through the DNA and the human spirit residing in the body, and creating the memories and the knowledge that is held in the human body. While a lot of this is stored in the human's physical brain, there's still a lot more stored in the bones of the human, within the crystals inside the bones. The bones can be regarded as the repository of much of the experience of the person during their lifetime.

Of course, the saints themselves led very blessed lives and their wonderful memories, knowledge and experiences were stored within the bones – we can use energy and knowledge interchangeably here. The primary place for the storage of the energy/knowledge was the bones. So medieval people would collect bones because they would be able to read out the energy from the bones during their ceremonies.

As you know many pilgrims went to places like Santiago de Compostela just to stand near the bones – they were prevented from touching them by the church authorities – but the bones were always placed in a location where people could pass very close to them. You have done that yourself, you have passed very close to the bones of Saint James in the cathedral of Santiago de Compostela. You know there was a special place under the altar where people could pass within a few centimetres of the bones and you have done that yourself. You didn't realize as you did it in this lifetime, but you were making a connection with your past lifetime when you were involved in securing those bones and bringing them to Santiago. (I remember that this act of passing close to the bones felt special – MKS.)

The pilgrims that came to these places wanted to be able to pass close by the bones, because then they could experience some of the energy that was stored in the bones. They could "read out" some of the energy that was stored in the bones and thereby receive a blessing from the knowledge that was stored in the bones. This was the main value of collecting bones and relics of the saints. This was so that the energy that the saints had accumulated could be passed on to their – we were going to call them admirers but let's stick to the word pilgrims. A definition of pilgrims could be: people that wanted to share in the energy of the saints.

They were also looked upon as vacationers, holidaymakers, because to go on a pilgrimage was a great adventure and involved much travel and seeing other parts of the world. So it was a very popular thing in medieval times for people to do. As you know that practice still continues to this present day, people walk across northern Spain from France and visit Santiago and share in that pilgrimage. However, for those modern pilgrims, the journey is a more important part of the ceremonies than the connection with the bones, because the value of the bones is not realized and the energy within them is not experienced to such a great degree, because the faith of the people in modern times is different to the faith that was held by the medieval pilgrims. We think medieval pilgrims got more from their experience of being near the bones than from the journey, whereas in modern times the journey, and the effort in making the journey, is the driving force for the pilgrimages that happen now.

The other aspect that we want to talk about is that while all humans store energies, knowledge in their bones in the majority of cases the information that is stored in the bones is not very great. Whereas in the case of the saints their lives were so blessed that much more energy / information was stored in their bones. We refer in the previous session like this to the example of Saint Teresa of Avila who led such a blessed life and who went into religious ecstasy on a regular basis – so much so that she levitated and was known as the flying nun. Her case is typical of many of the saints who led such blessed lives that the energy they absorbed into their bones was much more than normal humans, and as a result that energy stayed in the bones after the passing of the spirit of the saint back to our realms. So the bodies of the saints, in particular Saint Theresa, did not decompose for many years, hundreds of years in the case of Saint Theresa. The bodies were preserved by the sheer energy that was stored in the bones. That energy allowed the body, minus the spirit, to continue to exist in a relatively stable physical form with no deterioration taking place for a long time. Eventually the body did deteriorate because the energy weakened or was removed. – We are told that the energy was removed from the bones after some great length of time after death, the energy was required for other lifetimes and was recalled by the spirit that put the energy into the bones in the first place. That is why the bodies eventually decomposed, as the energy was taken back into other realms. But as a demonstration of the energy that was in the bones, the bodies did not decompose for several hundred years. Church people observed these things, marveled at them and regarded it as a miracle. It is a miracle, but you, Malcolm, are beginning to learn the mechanisms behind the miracles and how those bodies can continue to exist without decomposition. That process is continuing with modern saints who have died – Mother Teresa is an example of that, her body will not decompose for some time. (Strange coincidence that her name is the same as the saint in Avila – is it not?)

Now moving on to another part, we want to talk about what you can do Malcolm in this respect. When you construct your torsional pendulum, we want you to do a number of tests. We know you already know about testing this pendulum with energy fields. You already tried holding paintings made by Michelle near your first attempt at a torsional pendulum. You will find when you construct a full-size one, as we described to you, that you will be able to use this pendulum to detect those fields, it will be very sensitive to them. One of the things we want you to test in this way will be some bones. You will detect some fields from simple animal bones because animals, of course, store their memories in their bones just as humans do. But the energies that are stored by animals are relatively simple and you will not get a very big reaction from the torsional pendulum. If you can get some human bones – and we know this is not easy, you may find some way in which a part of a human skeleton can be made available maybe through a medical situation, you know that you have held a human skull during a first aid lecture, so there are human bones available for testing with the torsional pendulum – you should get a very big reaction to those. This is something we have to work on together, finding suitable human relics for your testing. It will lend great weight to your experimental results if you can show the much bigger deflection of the torsional pendulum from human bones compared to animal bones compared to inanimate objects. It will be the biggest effect you will see. Of course if it had been possible for you to test a bone of one of the saints, then you would get an even bigger reaction from the torsional pendulum in that case, because of the much greater energy stored within the saint's bones. You would be measuring the information/energy that came from the saint's bones and you would see that it was at a very high value. That's what it was that the pilgrims were experiencing when they went on a pilgrimage and passed the saint's relics in the cathedral.

Well Malcolm, we think that is enough for today. That is very specialized part of the story of crystals and their impact, how they are used to store energy. Spiritually, it is a very important part, that is why we bring it to you first, because this will set the scene for our book number two. The energy of saints is stored in their bones and can be measured still, that is the basic message that we want you to tell as one of the chapters in our book. We are not sure which chapter this will be, but as before we will arrange the material in a suitable sequence after you have written the chapters. We just want you to write the chapters more or less in isolation, although of course there is some connection between the chapters as there is in our book one. In book two, you will be writing about one of the aspects in each chapter and you will refer to other chapters because there are always connections between them, as you have seen in book one. This is your first topic that we want you to write about and to absorb. Of course, you will need to do some research as you understood as we have gone through this; you need to look up the crystal structure of bones and we know you can do that from your past experience.

We feel you have enough for this section today. It's a very important starting point for your work on book two. Something you can really get your teeth into – which is kind of a joke because the teeth are very like the bones, they have got that material stored within them as well. That's an important experimental point that maybe you can find human teeth more readily than human bones. You should test teeth in the torsional pendulum because teeth have as much energy absorbed in their crystal structure as bones do. That is a useful footnote to our session today.

So with that Malcolm, we thank you for doing this once more.

We know you enjoy doing this and we certainly enjoy passing the information to you. We are very thankful that you are able to do this for us and bring this information into the world. We know you will have fun doing the research and writing about this. So we leave that with you with our thanks and blessings. All the blessings that are ours to give we give to you. We thank you for doing this and we wish you a very happy day.

Crystal Light.

Time elapsed 38 minutes.

Malcolm K. Smith / August 9, 2010

Figure 1 – 1
A simulation of atoms in a grid inside a salt crystal

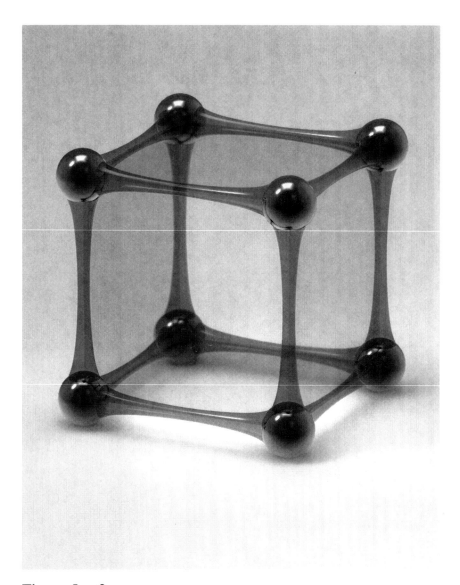

Figure 1 – 2
A simulation of a cubic unit cell inside a salt crystal

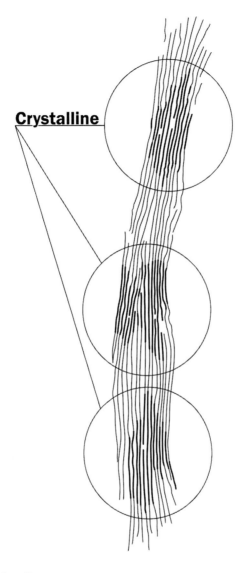

Figure 2 – 1
Polymer molecules come into alignment to form crystals in a semi-crystalline material

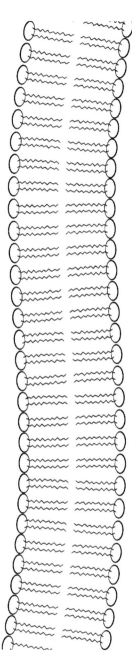

Figure 4 – 1
Lipid molecules form a double layer which acts as a liquid crystal

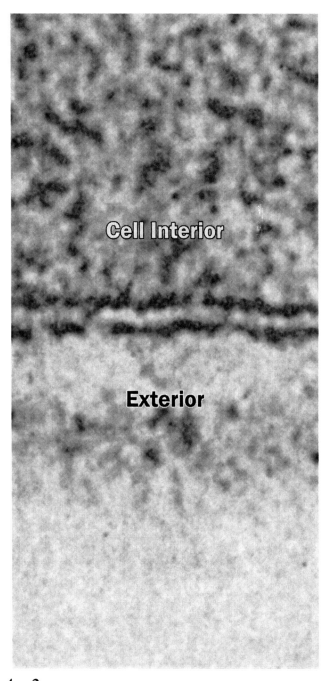

Figure 4 – 2
Microscopic view of a cross-section through a cell wall showing the double lipid layer

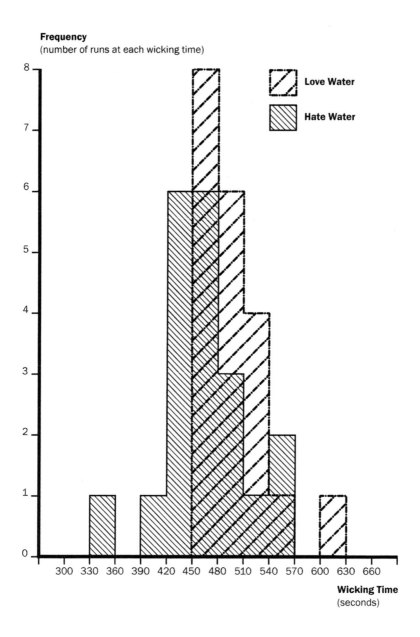

Figure A – 1
Wicking time frequency diagram

Appendix 4:
Angel Communication, August 16, 2010

Hello again Malcolm! Here we are in the nice cool basement of your house on this hot August day. We are ready to talk to you again about crystals.

Recently while you were moving some fossils around, the thought came to you that in the rock that constitutes the fossils, there must be some crystals. From that you continued on to think that crystals contain information – maybe the information in the crystals in the fossils can tell us something about the world at the time the fossils were living things on the surface of the Earth.

You were quite right – incidentally that thought was put into your mind by us because earlier today you wondered if it was time for us to have another session of channeling. We knew you were ready to do this, so when you were moving the fossils around in preparation for a children's workshop on dinosaurs we saw an ideal opportunity to give you a clue as to the information that we would be covering in the next channeling session. We didn't think you would make it as fast as this but obviously you were excited by this concept and that was enough to get you back on the pendulum asking if we could have another channeling session. We are delighted to give this information to you, we have had it ready for you for some time.

Now today we are going to go back to our original session, when we talked about crystals being able to store information/energy. Just to recap, you remember we said that crystals can be considered as frames of energy into which pictures made of energy could be placed and stored. We gave you an analogy of music notation in which the unit cells of the crystal could be replaced by a music staff and bar lines. Then for each measure of the music – that's the space between two bar lines on one staff – you have a two-dimensional version of a crystal's three-dimensional unit cell. Into that two-dimensional musical space you put concepts, in the form of musical notes and rests, which tell your musicians how to play that particular section of the music.

By analogy with that, we can say the unit cell of a crystal is like a measure of music and into the unit cell we can put information/energy, just as you put in your music measure notes and rests. Those notes or concepts that are stored in the cell can be read and they can be removed. So this is a very useful storage and reading device that we have here in any crystal. But of course, most humans do not realize that inside any crystal there are billions of these little storage frames that can hold a vast amount of information.

We have talked to you about how in crystals of apatite in bone structure, information is stored that was put there by the creature that formed the bones – information about the creature's life. If the creature was a human then there was information in the unit cells – in the musical measures – of the crystals, concerned with the lifetime of that person. If the person was a saint, then it was very exalted energy, a very high level of energy, associated with the life of a saint. That is why in medieval times churches collected the bones of saints and stored them, so that people could have the benefit of passing close to those bones and reading out the energy. Humans and some animals are able to read out the energy of the material that is stored in the crystals without removing the energy from the crystals. They can read the energy and gain information from it. It's like reading a book in a library – you don't have to remove the book from the library. You can go to the library, take the book off the shelf, read the information that is stored in the book and put it back on the shelf.

That is an analogy of what humans are doing when they read the energy that is in a saint's bones when they go on a pilgrimage.

For example, we cited the case of Santiago de Compostela in Northwestern Spain, where in the cathedral there the bones of Saint James are stored. We told you the story of how you were involved in the recovery of Saint James's bones and their installation in the cathedral. Also, we told you how in this life you passed within centimetres of Saint James's bones when you visited the cathedral of Santiago de Compostela. You didn't know what it was that you were so close to at the time, but you had the feeling that there was something special about this experience. You detected some new influence coming into your life and we tell you now that it was the influence of Saint James. It was from the energy stored in his bones that were within a few centimetres of where you were standing. Those bones could be read by you, you perceived the special energy that came from them, but you didn't know at the time what it meant or what you had to do with it. We tell you now that was a very important event in your life, because it awakened within you some knowledge that had been latent within you about how you helped bring the bones to Santiago de Compostela. It also awakened within you a sense of having to do something with this information. It was in the 1990s that happened and you did not know what it was at the time.

(I remember in the hotel in Santiago de Compostela we stayed in a room on the top floor. I woke up in the middle of one of the three nights we stayed there and saw the constellation of Orion – which was significant for me – perfectly framed in the skylight in the ceiling. I "knew" then that something special was happening – MKS.)

Since then you have taken on this work of writing books and explaining all these physical effects to your fellow humans. Now when you look back it makes sense. That was the time that the knowledge was brought to your attention and the process was started that led to you sitting here talking for us today.

Just as the energy of the saints can influence people like you,

Malcolm, so it influenced many pilgrims that believed in these possibilities, and those beliefs were very strong, particularly in medieval times. As we told you last time, the belief system of Christian people has changed somewhat. Now their energy goes more into making the pilgrimage rather than the end result of being close to the saint's bones. But that is because the – "magic" if you like – the special feeling of the energy from the saint's bones, and its impact on people, has been forgotten since medieval times when it was so strong. Now all people can see is the preparation for the somewhat difficult task of walking, or cycling, all that distance from one corner to another corner of Spain and in making the pilgrimage. We want you to write about these things, so that people see there is a reason for going to places like Santiago de Compostela and they will rediscover the wonderful information they can read out from the saint's relics.

Now we change to our original topic that we told you about at the beginning of this channeling. That is if you can read out energy about saints, then you can read out energy about the life processes of any creature, and the conditions around the person or animal at the time the information was laid down in the bones during the lifetime of the creature, person or animal. This means now that by looking at crystals of apatite in animal bones, we can find out about the conditions of life of the animal while it was moving about on the surface of the Earth. By a similar token we can go back in time and by reading out the information in crystals in the rock of fossils – the fossils are composed of rock which is composed of crystals – so in the crystals in the rock will be the conditions that were stored as the creature or plant was fossilized. So we have a record in time of the conditions that were existing a long time ago when the fossils were formed and when the creature that is fossilized was traveling about the Earth.

The question is how do we read out that information and learn about the conditions that existed on Earth during the life of the animal? How do we read out that information so that we can study conditions on Earth during prehistoric times? A similar problem which is not so difficult is to read out conditions during historic – not prehistoric – times. For example, if we find the bones of some historic person, not necessarily a saint, can we read out the conditions that existed during that person's lifetime? But how do we do that is the question? We are sure you are concerned about that, being a scientist. We assure you there are techniques for reading out that information and you, Malcolm, will help develop those techniques, so that the information can be reached.

To reach that information we have to make use of the cosmic lattice. Because the cosmic lattice runs through the crystals and is, if you like, the conveyor that brings the information into the crystal in the first place where it can be stored. The cosmic lattice is the means by which that information can be removed from the crystal if that's necessary – but of course we don't want to do that – we just want to treat the crystal as a book in a library and read the information. Once again the cosmic lattice is the means by which we read out the information stored in any crystals such as in bones and fossils.

The technique becomes more manageable now because we have another process – which we have talked about and you have written about in our book – which is analogous to that. That is the process by which you create artistic things, artistic objects and artistic measurements of sound and time that you call music. If you remember from our book – Chapter 7 – you talk about creativity using the Golden Angel of each person, or the higher self, to gain access to particular energy resonating in the cosmic lattice. We can use a similar process whereby your Golden Angel can access the energy that is vibrating in the cosmic lattice within the crystals of interest.

So the process you would follow would be essentially to connect with your higher self, your Golden Angel, and ask to have information relayed to you by the Golden Angel, information coming from that stored in the crystal using the cosmic lattice as the carrier wave – if you like.

(As I write this channeling, and the recent others, I keep seeing childhood memories. This experience makes me realize that maybe our long-term memories are stored in our skull, literally in the bone of our skull! My Guides said: ***Bingo! You have made another big connection.*** I then realized the significance of the crystal skulls they said: ***Eggsactly, that is the significance of the crystal skulls.*** I presume for racial memories. MKS.)

It is a similar process to creating a painting or an invention, which in your case is perhaps more familiar, Malcolm, where you have to give a great deal of thought, or intent, around the information you require. You study what you can find out about that and then you relax and let your mind follow your Golden Angel to the relevant part of the cosmic lattice. Your Golden Angel will help you by taking your spirit into that part of the cosmic lattice where it can resonate with the cosmic lattice part of the energy that is stored in the crystal. When your spirit resonates with the energy stored in the crystal, then that information will be brought back by your spirit and will be relayed into your body. Then you will have that familiar feeling of energy coming into your body from somewhere other than you, just as artistic inspiration comes to you from the cosmic lattice in the same way.

Alternatively you can train yourself to resonate with the cosmic lattice at a particular wavelength of thought and connect with the cosmic lattice yourself. After all this is what you do day to day as you create your reality. But that is more difficult, because you do that part unconsciously. All you know is that in order to access good things from the cosmic lattice, you have to think good thoughts around the thing that is desired. This is the basis of the law of attraction. You can do it that way, but it is more difficult to control the process. Although some talented people can do it and that is the basis of the skill called psychometry.

This is the skill of people reading out from objects details of the person who owned the object. For example, a police psychic may be given a garment of a person who is believed to be dead after some foul play. By holding the garment the psychic can tune into the vibrations in the tiny crystals in the material of the garment and read out from them impressions of the person at the moment of their death. That is something that is possible, but only by talented people who have that special ability of psychometry.

We think it is simpler and more achievable if you follow the conscious exploration of the energy in the crystals of interest by enlisting the help of your Golden Angel to show you the way to the right part of the cosmic lattice that is within the unit cells of interest in the crystal. That is the process we recommend you follow and relate to other people as being more workable, more realizable, by a large number of people. Not everybody will be able to do it for a start. You have to believe in your Golden Angel and have a good relationship with your Angel. Of course you do, Malcolm, and probably many of the people who will be reading our book will be in the same position. But of course there are many people who don't believe in Angels. They will have great difficulty getting to any information like this.

That is one of the benefits of living a spiritual life. Your senses are greatly magnified and you are able to pick up – not only things that are happening in your world at the present time – but also by going back into the past you can pick up information about the situations and events in past lifetimes. That is stored in crystals within the fabric of living creatures: people, animals and plants and in their skeletons and fossils. It is apparently unavailable to anyone, but we tell you it is available and can be read just like a book.

The answer will not come out in words, it will come out in impressions, for example if you explore the bones of a tiger for the memories, the information stored within the cells, then you will see impressions of life in the jungle somewhere in Asia where the tiger lived during its lifetime. It is an acquired skill to interpret these impressions, to put them into words to convey the feelings the animal had. It's easier to work with bones of humans because the concepts, the information that is stored in the bones, are more familiar to human thought, closer to human thought and it's easier to resonate with that structure. It's easier for the human who is reading the library book that is representing the unit cells of the crystal to connect with the thoughts and impressions of the human in whose bones a record has been made.

That is the process that we tell you can be developed. It has been known in the past in previous civilizations and it can be rediscovered and reestablished. Maybe you need the help of some talented psychic people who are able to use the second unconscious method of reading directly memories stored in objects – that maybe is a way for you to go – to get training in how to connect with it. But we think if you connect with your Golden Angel Mikael he will be able to help you find your way to the right part of the cosmic lattice that is within the crystals by focusing on the crystals within the bones. Then you will be brought to the right part of the cosmic lattice where you can pick up memories.

You detect hanging around at the back of our consciousness another related effect. We can see that you are getting excited about this next concept we want to bring to you. We will finish with the crystals within bones and fossils and consider that complete now.

Now we turn to another type of crystal which exists in bodies, that is the liquid crystal which forms cell walls. In Bruce Lipton's book *Biology of Belief* there is an ideal representation of the material that forms the biological cell walls. Lipid molecules – molecules shaped like a stick with a ball on one end, you know what we mean by that, like a surfactant molecule – those lipid molecules are all lined up like soldiers on parade. But these are soldiers that are marching so these crystals are liquid. They are crystals that can move if necessary. But they still have unit cells and they are still able to store memories within those unit cells, just like the unit cells of crystals in solids.

This means that within any human, memories are stored not only in bones but also in other organs of the person. We think you can see where this is going. It means that within any particular organ of the body memories are stored within the cell walls. Bruce Lipton was very ahead of his time in thinking about the cell walls being such a repository of information, although of course he arrived at that conclusion from his study of cell biology. We're telling you – in coming from the direction where we are – where the liquid crystals store the energy. This means that the cell walls have a great deal of information/energy stored within them. And it's not only the cell walls. There are other parts of the cells, in particular the alpha helices – the double helix – of DNA which of course has got a great deal of information stored in its crystals and in its chemical structure as well. Around the DNA are the ten magnetic strands and within those strands there are vast amounts of energy storage. People can change their DNA, the information that they brought to Earth, by changing their DNA storage. You've done some of that yourself Malcolm, you know what we are talking about.

Coming back to the implications for the energy storage within liquid crystals in cell walls, this means that when an organ is transplanted from one human body to another, then the information that is stored within the cell walls is transported to the other person. This explains the strange events that are occurring more and more in human hospital lore in which people receive donor organs, but not only do they receive the organ itself, but also memories of the person that formed the organ in the first place, the person that grew the organ in their body before it was donated. This explains the sudden changes in personality that sometimes occur. It's not very common, very often the transplanted organ doesn't have any effect. But there are some surprising cases Bruce Lipton writes about in his book and it seems he thinks something like this is going on. There are some remarkable cases, for example a very gentle woman who has had a quiet life may suddenly, on receiving a donated organ, acquire a taste for riding motor cycles and smoking cigarettes. This is because the organ that she received was constructed by a person who had those tastes. It seems that a sudden change in personality accompanies the organ when it is transferred from one body to another. In a way that's true, but it is the result of the energy that is stored within the liquid crystals that form the cell walls. So here is another medical mystery that is being explained by this understanding we are giving you today.

It might be easier to relate to the information in liquid crystals than to relate to the information in normal human bones. The easiest information to relate to for a human is that stored in the bones of a saint, simply because the saint had such strong beliefs and faith that their energy was exalted, so it has a big impact on humans. That is the easiest one to relate to and you experienced that when you were standing near the bones of Saint James in the cathedral of Santiago de Compostela. If you want to practice such techniques you need to immerse yourself in the knowledge that comes from holy places where saint's relics are stored.

What we recommend that you finish up with in this particular channeling session is that we recommend, as we did before, that you obtain human bones or teeth – teeth are a good source of apatite crystals as well, you will have equal success with teeth as bones – and try to explore the energy stored within those ordinary human remains, bones or teeth, with the help of your Golden Angel Mikael. It will take some practice, because this is a new technique, but you will be surprised at some of the results you obtain. We Crystal Light will support you in this and help you and direct your research to gain access to the information/energy stored within the bones.

We leave you with a picture in your mind of the Shakespearean play *Hamlet*, when one of the characters – we believe it is Hamlet himself – is holding a skull and saying, "Alas poor Yorick I knew him well." In that case, Hamlet was relating to the energy stored within the crystals in the skull of Yorick. That's the sort of thing we want you to do Malcolm. It's another field of study that you can explain in our book.

Now we think you have enough material to start considering writing a chapter about it. There will be a whole chapter on crystals within bones. Maybe it's best if you start with a section – not a complete chapter – on crystals from the scientific point of view, explaining what unit cells are, how atoms are lined up in a crystal and the different forms of crystals. You can relate this to crystals in the form of rocks and minerals and qualities of those minerals, how they can be used in certain ways. You have a book that is very good that covers the use of different minerals in healing situations. It all depends on the energy/information which is stored in the crystals of those minerals. If you write a chapter about that, it will set the scene for your next chapter in which you talk about the energy that is stored within the apatite crystals in bones. Then you can go on to discuss the energy stored in liquid crystals and the case of organs that are transplanted, apparently changing the personality of the recipient of the donated organ.

We detect that you, Malcolm, think this is a wonderful combination of spiritual information and science. We know you will enjoy writing about this. We are very happy that you are so anxious to explore these things with us and you will be happy to write about these great thoughts and tell your fellow humans about the wonders that await them.

We think that is all for today, Malcolm. Once again we thank you for doing this. We know you enjoy it and love your work, we are so pleased that you do. It's so pleasing that you are anxious to learn more about these things and you look forward to the next channeling session. We think that is a great quality that you have

to exploit Malcolm. We are very grateful to you for putting your energies into bringing this information to Earth at this time. With that with our thanks and our blessings which we bring to you we have finished today. All the blessings that are ours to give and our love we bring to you. We say goodbye for now.
Crystal Light.

This is the end of the channeling for August 16, 2010. The time elapsed is about 48 minutes yet while I am channeling it seems like only 10 or 15 minutes. I guess when I'm channeling I'm running on Angel time. I am very honoured to be able to bring this information to Earth for Crystal Light and all my Guides and I thank them very sincerely. Thank you God, All That Is for allowing me to bring this information to my fellow humans – MKS.

Malcolm K. Smith / September 2, 2010

Appendix 5:
Angel Communication, October 29, 2010

OK Malcolm, you are all ready to go now. Thank you for setting this up quickly. It's kind of a surprise we know, but you've been having some warnings that this was coming over the last couple of days. Basically, we have come today to answer your questions about the shift in the dimensions, about the movement of so many humans to the new Earth and how this impinges on your work in writing the books.

First of all we want to confirm very strongly that the shift of the dimensions to the new Earth is about to occur. It could happen in the next year but most likely it will be happening in 2012. We can't pin it down more precisely than that because there are various factors involved that we can't go into now, but believe us it is an event that is inescapable. It is definitely going to happen, the only remaining question about it is precisely when it will happen. We know it can't be later than December 2012, which was the time you were given, and we think it's very likely to happen in the six months before that time. That puts it somewhere between June and December of 2012. That is our definite time slot that we have been given and we understand that is when the shift will occur.

All the humans of good intent will make the move from the four dimensional existence they are in at the present time to a new five dimensional existence on a copy of the Earth – a new version of planet Earth which will be circulating around the sun added as a tenth planet to the nine that already exist. That is predicted in the crop formation (Alton Barnes, Wiltshire formation, July 2010 – MKS.) that you saw in the picture this year – the ten planets that will be in the solar system after the shift has occurred. In fact, the planet of the new Earth is already in existence and some astute astronomers are noticing there is something on the opposite side of the sun to where the present Earth is. Although it is not completely visible, they are aware there is some presence and you, Malcolm, have seen a couple of references to the existence of another planet. This is the start of how the process becomes physical – the manifestation of a planet.

You, Malcolm, have traveled to that planet in dreams – in fact the birthday party dream you were just writing about takes place on that new Earth. Another of your dreams a few days ago, of driving along a very peaceful lakeshore road, which reminded you of Switzerland was also on the new Earth. You, Malcolm, and many other people who are immersed in studies of this kind, go to the new Earth during your sleep periods. That is why you sleep so soundly recently, it is because you are out-of-body visiting that new Earth and preparing to exist on that planet permanently. We are sure you are going to love the existence on the new Earth and all these travels that you do are in anticipation of the move and the life that you will enjoy on that planet.

This brings us to the topic of writing our books. We know that you have had a good start in writing about the energy in physical objects, in particular crystals, and you have already prepared a draft of the first few pages of that book. Then you heard about the shift to the new Earth, about half the human race moving to this new planet and you lost heart, your momentum in writing. We would like to tell you that you should continue writing as if you were going to live for many more years on this present Earth where you are now. In fact you will be living on the new Earth within two years from now – we guarantee that. We can't say more precisely how soon but within two years of this December you will definitely be living on the new Earth. All those around you and dear to you will be there with you.

You will continue your work of writing. The physical facts and the spiritual facts that you write about in our books will apply on the new Earth just as much as they did on the old Earth. But there will be additional factors that you are only aware of in out-of-body states, such as the presence of your Guides and being much closer to your Guides and to us Angels as we help you. On the new Earth, you will be able to contact us very strongly and even see our presence. So that will change the authenticity of the things you write about – the facts that you are given on the new Earth will become much more authentic. But that does not mean that you should stop writing now on the old Earth, because this is a valuable point of view that you have at the present time. All these things about information/energy being stored in crystals are new and revolutionary. In those books – the one you have already written and the one you are about to write – you will have valuable insights as to how physical objects can hold energy which is vibrating at certain frequencies.

The fact that is very difficult for people to pick up those energy vibration frequencies on the old Earth is a valuable lesson, because it shows how the interaction of humans with their physical surroundings takes place and the potential that's there for much closer interaction – a more precise perception. Because it's so difficult for humans to perceive energy vibrating in physical objects it makes the information that you are gaining through us much more valuable. You will have a unique view of these things as a result of writing about them on the old Earth. When you move to the new Earth, this sort of thing will be much more commonplace because of the closer contact with us and your Guides. That means there won't be such novelty value and many people will be able to make those contacts, but the understanding you achieve in the "school laboratory" of the old Earth means that you will have a really deep knowledge of how these things work. For that reason, this is a valuable tool for humanity, if you do this work now you will understand in detail – after quite a struggle to put all these things into words, and we appreciate that it is not easy for you to do that – but as a result of your enlightenment that you gain from studying these related topics and putting it all together you will find, when you get to the new Earth, that you will have a really good understanding enshrined in our book. So it's well worth working away for the next two years to get everything down on paper so that in the new Earth you have that information at your fingertips and you are able to explain, so that everyone understands, how those skills, which are fairly commonplace in the new Earth, are achieved.

Do not think it's a waste of time writing books now, because every page that you write will be faithfully transported to the new Earth. You will have all our books available to distribute among other humans when you are in the new Earth situation, in fact you will become one of the main explainers of the transition that has occurred.

That brings us to the next topic we want to put before you. We've dealt with the energy in objects and the importance of writing the book now so that you will have all this information in the new Earth situation. But in addition to that book and your present *Spiritual Chemistry* book – and all the things you said in that book will become much more obvious when you are in the situation of the new Earth – in addition to those two books which we want you to have available at the time of the shift, we also will need you to write more books in the new Earth situation because then people will be confused about where it is they have ended up. Most of the people that will be making the shift will be completely unaware that they are about to move in this way. When they find themselves on a new planet which is very beautiful, very peaceful, very calm, very loving they will be extremely happy that they find themselves in this situation but most of them will be confused about how they got there. They will wonder what it is that has happened and why it is that many of the people they knew, maybe some members of their family, are not with them on the new planet.

We wanted you to read that book *Left Behind*, because that situation will occur on the old Earth, where people will disappear and the people who are left behind will be puzzled as to what it is that has happened, where so many people have gone. You will be able to help those people as well by transferring back information. But the main ones we want you to be concerned with, Malcolm, are the ones who have made the transition to the new Earth and who find themselves in this beautiful place although without some members of their family and friends. You will be needed to explain what has happened to those people and you will have the help of people around you who are light workers, like Ann Perrick and Michelle DeMello, they will be able to help explain. This is a very important task that you will be facing when you get to the new Earth, and we want you to have these two books completed and available to anyone that wants to read about how things work on the old Earth. Then it will make sense to people how they have arrived on the new Earth and why they are in these wonderful calm, peaceful, loving surroundings.

That was the main message that we wanted to convey to you today because we've noticed that you have been a little confused about what your role was in this changeover and how your work of writing books was affected by the potential of this change. Understandably, you were feeling confused to a certain extent and also you felt it wasn't worth writing if those books you produced were left on the old Earth and you wouldn't have them on the new Earth. But we assure you that all our books, all books of good intent – like people of good intent – will be transferred to the new Earth and every word that you have written on the old Earth will be available to you on the new Earth. Not only will they be available, but they will be fairly essential that you have that information at your fingertips to help the people who are recently arrived on the new Earth and have no idea how they got there and where they are. You will be able to explain all those things to them. So that is the task that we lay before you today, Malcolm, and we know you will be much more settled and relaxed in doing your work now that you understand this change.

As always we thank you for helping us with this work, for cooperating with us and being so ready to put down our words on tape and then transfer them to paper. We appreciate your help Malcolm, and we love you and we give you all the blessings that are ours to give. Now we say goodbye and once again thank you.

Malcolm K. Smith / November 5, 2010

Appendix 6:
Angel Communication, November 6, 2010

Well Malcolm, glad you were able to fit this in and find a tape on which to record this message. We have been watching you in your research, and you are understanding a new principle, which is not so new in the sense that it's been written about before (*The Field* by McTaggart), and the people who are doing research on the imprinting of water and the homeopathic remedies already know about this effect. We want to make it more general than that. We connected with your Guides and decided to give you a "heads up" as you call it on how this all works together. You, and your Guides, are quite right in the assertion that all molecules give off electromagnetic signals. This should not be surprising since molecules consist of atoms bound together by electrons, and those electrons vibrate at characteristic frequencies – each bond has a characteristic frequency of vibration – so when you put all those characteristic vibrations together, each type of molecule has a unique vibration signature. For example, carbon dioxide molecules have a different vibration signature to that of water molecules, but every carbon dioxide molecule has the same vibration signature as all other carbon dioxide molecules. The signature is not characteristic of individual molecules, so much as characteristic of a particular type of molecule.

Those molecular signatures are used in space research, for example to detect water on the surface of another planet. Scientists can detect molecules on the surface of any planet simply by tuning in to the characteristic molecular vibration patterns in the light reflected back from the planet to Earth. This is the science of spectrometry which looks at the amount of light coming from the source – e.g. gas molecules in a planet's atmosphere – at each wavelength. With this technique each type of molecule has characteristic peaks in the curve of intensity of light emitted by the molecules when it is plotted against wavelength (wavelength of light is the scientific measure of color of the light). For example, earth satellites can detect the fall colors of trees in the northern hemisphere by a change in the relative amounts of red light and green light.

That is very useful over all the universe, because this is a way of signaling over great distances the presence of certain molecules. From that, humans can deduce what is present in other planets, in suns and stars by simply looking at the light that comes from them. (For example, helium was first discovered on the sun by its spectral signature that had never been seen before on Earth – MKS.) This should not be a surprise to anyone like you, Malcolm, because you have studied the light absorption characteristics of molecules and you studied how certain bonds in and between molecules changed the light absorption spectrum of each of the dye molecules you were working on. For example, if the dye molecules were hydrogen bonded together in the solid state then they had a different light absorption spectrum from the molecules that were dissolved in alcohol. Those subtle changes in the light absorption spectra corresponded to changes in the hydrogen bonding between the molecules. So it's not surprising when you look at the reverse of this light absorption process to find that molecules can also send out light vibrations and their electromagnetic spectrum signature reveals their structure.

What is surprising – from the book *The Field* that you were reading – is that certain molecules, in particular water molecules, are able to record electromagnetic signals. As you read, water can act like a tape recorder. That means that certain clusters of water molecules come together in response to particular electromagnetic signals from biological molecules that have been present in the water. Those signals that are stored in the water can act as a substitute for the actual biological molecule. Once a molecule has been in the water and a signal has been imprinted on the water molecules, then that signal imprint remains with the water molecules even if you take away the original imprinting biological molecule.

An analogy that comes to mind is that this is like crop circles. The energy stamp that is put on the crop makes the crop stalks lie down in a particular pattern and that pattern is permanent until the farmer harvests the crop. In the same way, biological molecules can put an energy stamp on water molecules that are around them, and the water molecules retain that characteristic energy, which is like a microscopic version of a crop formation pattern, except that water molecules hold the energy stamp image instead of crop stalks. Then you can take away the biological molecules and the water molecules still hold the energy stamp as if the biological molecules are still present. This is the basis of homeopathy. Homeopathic medicine depends on this effect. The biological molecules that are involved in the cure are no longer needed to be present, instead their energy stamp has been put on the water and even although the solution is diluted many times, the water retains that energy stamp just like stalks of wheat retain the energy stamp message in the crop formation. Yes, we think that is a really neat analogy.

That explains homeopathy and all those very controversial results that were obtained by people like Benaviste as recorded in the book, *The Field*. Now another aspect of this is that the electromagnetic signals from molecules can be recorded on other substrates such as very delicate computer-based recording materials, and those recordings can be sent from one place to another on the Earth. That means that the energy stamp is the information that is recorded and sent to another location. So in this case that we are talking about, the water recording the energy stamp of the biological molecules, is now replaced by a synthetic recording medium which is very sensitive and is able to record the signature vibration of the biological molecules. That recorded vibration can be sent to another part of the world and can be used to influence other molecules just as if the original biological molecule was present.

However, there is another factor that we must include in this account; humans can influence those recordings. That is the basis of the story in the book about a female laboratory worker whose body had a strong influence which disrupted the homeopathic process when she handled the solutions. The process did not work because some aspect of the electronic signal from her body was very strong and it neutralized all the recorded signals from the biological molecules. An analogy to this we will mention here is that you record this message on magnetic tape, Malcolm, but if you have in your pocket a very strong magnet, it can interfere with the recording process and prevent the message from being recorded. That situation occurred with the female laboratory worker who stopped the homeopathic experiments from succeeding. It shows that the energy stamping of water by biological molecules is a relatively delicate process and can be interfered with by certain humans who have a very strong electromagnetic field around them. You were thinking about a person you knew who was able to pour boiling water into a wasp nest and not one of the wasps would land on her body and sting her, although they had already stung you six times. This because that person has a very strong electromagnetic field around her which repels the wasps and it is similar in a way to the female laboratory worker who repelled the effect of the biological molecules she was working on. So all these systems are quite explainable in terms of recordings and crop formations.

Now we move on to Dr. Emoto and the process by which he influenced water molecules to produce certain ice crystal types starts to become apparent. For example, water that was polluted had an energy stamp of the pollutant in the water so by putting a label that said "love" on the container they would have influenced the water's recording of the pollutant. He was able to neutralize the water's recording of the pollutant by his intent. Similarly he could take fresh water and energy stamp a message of love on the water by his intent. That intent can be extended to include beautiful music or unharmonious noise, as you did in your experiments. In the case of beautiful music, the energy stamp of the water in containers labeled "love" would be energy of the music which has a positive effect in biological systems. The water treated by discordant noise – that you labeled "hate" – would have a different energy stamp.

The recordings in the water molecules of those two different energies are quite different and influence the physical properties of the water. The "love"-treated water migrates at a different rate up the strip of filter paper compared to the water from the "hate" sample. What is happening there, is there is a different structure put into the water by the love or the hate signal, and that is like two different crop formations. The size of the clusters of water molecules depends on the recorded signal. In the case of the "love" signal the signal impressed on the water molecules results in small clusters of water molecules – in the analogy equivalent to small crop circles. In the case of the "hate" signal recorded in the water the clusters of water molecules are very large – in the analogy equivalent to large crop circles. When the water flows through biological systems such as the very small openings in cell walls the apparent viscosity of the water depends on the size of the clusters. The large molecular clusters of the "hate" water find it very difficult to get through those holes compared to the "love" water which has a lower viscosity owing to the small clusters of water molecules. The lower viscosity "love" water flows much more easily through the openings and this is beneficial for the health of cell and the whole structure – be it human, animal or plant.

Now we come to the main basis of your studies this morning. The question you originally posed was: is there any evidence to support the idea that certain crystals have certain biological effects? For example, you were thinking of the beneficial effect of the black tourmaline crystals from which you made a necklace for your relative who was fighting breast cancer. What is the basis of the power of that black tourmaline crystal?

It is all related to the idea of energy being emitted in the form of an electromagnetic signal by the molecules of the material from which the crystal is formed. In the case of the black tourmaline crystals used in the necklace, the molecules of a complex borosilicate emitted a particular electronic signal, and that signal was active in helping to change the biological system and overcome the cancer. That is how all the crystals work.

The recommendations that are listed in encyclopedia of crystals like the one you possess are all based on time tested experience, traditional results that have been gathered over many centuries of observation. It is entirely empirical knowledge, but revealed underneath, it is a scientifically-based system in which the chemicals, from which the crystals are made, emit an electronic signal which has been found to have a beneficial effect on the person wearing the crystals. In most cases, the effect of the signal is imprinted on the water in the cells of the person's body – so you could say that the crystal is acting like the biological molecules that are imprinting the water in a homeopathic remedy. That water in the cells of the person being treated is transferring the signal to the biological systems and giving a beneficial result.

In some cases, the electromagnetic signal from the molecules in the crystal is directly influencing the medical situation of the person wearing the crystal. But in most cases, the signal is transmitted through the water as we described, in which the water records the signal and acts in the biological system as imprinted water, which works like a highly diluted homeopathic remedy.

There you have an explanation of how crystals influence peoples' health. It's all connected with electromagnetic recordings in water. In almost all cases, the water acts as the intermediary, because there is so much water in human biological systems. Wearing certain crystals on the body is like creating a homeopathic remedy for a particular ailment, although it may puzzle a person as to why they are recommended to wear those particular crystals. It is a beautiful system that works very well and does not involve any invasive action on the wearer. As in all these things, if the wearer of the crystals truly believes that they will help, then added to the equation of the homeopathic remedy is the further effect of the person's intent. Another way of saying that, is it gives the person a "placebo" reason to believe that the treatment will be successful. That placebo effect is added to the homeopathic mechanism and results in remarkable changes in the health of the person wearing the crystals.

We could say that there are a lot of "old wives tales" around healing by crystals, and that would be true, because that is how the understanding developed. The nuggets of information about what a particular crystal can do for a particular health problem have been gathered over the centuries, creating a body of empirical knowledge that has a basis in fact.

That is all we wanted to tell you today, Malcolm. Thank you for making this quick recording a possibility, we think you will find it really helpful in writing your chapter about energy in crystals. Of course, it will come back again when you write your chapter about energy in water. We thank you for speaking on our behalf and for recording this. We hope you will enjoy using this information in our book that you have begun and into which you are putting this new understanding. Some related work is being done by scientists, but they rarely apply it to the physical situation of the empirical knowledge that so many people hold. There's a tendency among your scientists to regard empirical knowledge as superstition, and the implication is that such knowledge doesn't work simply because they can't produce the effects in the laboratory. They haven't refined their experiments sufficiently to be able to detect these effects. We're glad to see that books like *The Field* are recording that some experiments are being done by a few individuals who are getting remarkable results, and a growing understanding is developing. You, Malcolm, can help that understanding to develop further by your simple explanations of apparently impossible experimental results.

We leave you with that now. Thank you for talking for us and recording this. We hope you have a happy day and we sign off with our love and blessings. Goodbye.
Elapsed time 40 minutes.

Malcolm K. Smith / December 6, 2010

Appendix 7:
Angel Communication, February 2, 2011

Hello Dear Malcolm, we are here again as always to help you with your work. Thank you for asking us to come and help you in your present state of being "stuck" and getting going on Book 2. You made a good introduction to the work and collected a lot of information, part of which we gave you, and that was all done in November and even earlier than November. But as a result of the preparations for your very successful workshop on January 22, 2011, that meant you lost the momentum that you required for the book. So it's fallen by the wayside shall we say, but that's no problem, we can soon pick up the pieces and get you started. All you need is some direction from us.

That is why we are here today, to get you going again. You will have added to that information we gave you, greater understanding now that you are living in the heart. We saw that you moved your emotional centre from your intellect into your heart centre and as a result you gained power. You just had a demonstration of that power last night when you allowed a psychic attack to come in and you dealt with and healed the effects that you experienced. This morning everything is back to the calm state it was before. A good night's work. Now you have added to your additional abilities of healing your mental states; this will help in your understanding that you will project out into Book 2.

This book is about crystals and the energy in them. There are lots of different aspects of crystals. For example, you had the experience on Saturday last, 29th January, of psychometry. For the very first time, you picked up details of Alison's experiences that were embedded in the crystals in the metal of the pendant that you were holding during the time you were doing the psychometry work. Those results were very accurate, and Alison was surprised to hear that this was the first time you had done psychometry. That was a good experience that you were able to channel that information, or do a psychometric reading, on the material in the pendant.

Most people would say the information was embedded in the pendant, but you know from your technical work that in the metal of that pendant are crystals, and it is within those crystals that the information is embedded. There is a lot of information that is very easily read once you realize that it is possible and you <u>trust</u> – which is the word that Judy gave you before you started out on that work. Now you have the experience of directly lifting the information from the crystals in a material of some object, which is a great advantage. In a way it's better that you delayed the writing of our book until you had that experience, because now you can relate to it on a personal level, in that you have done the actual reading of information from a crystal.

Now that you have done that, you will be able to do it many more times and you will become very proficient at reading the energy in crystals. Of course when you read the energy it's like reading a book; you don't take the energy out of the crystal, instead you read the energy which stays in the crystal so that other people can come and read it too. It's just like reading a book in a library, you don't need to take the book out of the library to be able to read it. You didn't take the energy out of the crystals – you just read it out. This is an important point to realize, that the energy will stay there for ever and can be read by any person that knows how to do it and has the trust to believe that they can do it.

You can do the same work with any other physical object and you can do it with liquids as well. In the case of liquids, it's not really crystalline structure – as you know from your chemistry –

but there are structures in water. Water is particularly good at retaining energy structures and allowing the energy to be read out again. That is the basis of homeopathy. We gave you much material on that topic and how it works, with the analogy of creating crop circles, in our channeling that you received on November 6, 2010, so we will not diverge into that area any more at this time.

We suggest that you start the first chapter of our book with your account of reading the pendant. What it was like. Put into words the experience you had and use that as an introduction that people will be able to relate to. In a way, the preparation that you have gone through and the delay that you have experienced all fit very well into leading up to this experience of psychometry. This is your keynote for the first chapter of Book 2. You can relate your personal experiences in that respect and collect other people's experiences as an introduction to the book.

Then, after that you can get into the details of how it works, starting with the structure of crystals – you have a lot of information about that – and that's where your intellect can come into play, in writing about the structure of crystals. But we want you to read all this material and remember these things through your heart centre so that it is not so intellectual. You have had a couple of guidance points from well-meaning beings; for one Darwin your Guide told you when you were out walking one day to try to be less intellectual in your approach to this work. Judy hinted at being less intellectual in the workshop on January 29. Both these people were helping you set the scene for the subjective description of psychometry and associated skills. You can refer to the colour work that you did at the workshop on January 29 and also the flower psychometry.

Really, it is just psychometry as well, but in the case of the flowers, you are relating to a living structure which is more sophisticated, we shall say. Let us diverge and talk about that.

You, Malcolm, were amazed when Wilma read so much accurate information about you from the single hyacinth that you brought to the workshop. You said to the assembled people that it was amazing because that flower had been in your possession for only about 15 hours before it was used, but all that information about your past appeared to have been downloaded into that flower. This is because the flower represented a pathway to the cosmic lattice. The cosmic lattice had all the information about your past and your characteristics already on there, and Wilma could have given a reading by connecting through your contact with the cosmic lattice. But because the exercise was to use the flowers, then Wilma focused on the connection between the flower and the cosmic lattice, which was existing ever since the flower first started to grow. As you know, Rupert Sheldrake has suggested that all plants, all living things, have their basis in morphic fields which are part of the cosmic lattice, so that hyacinth flower was very connected with the cosmic lattice. That allowed Wilma, or any person that is trusting and open-minded, to connect with the flower and to travel via the flower's connection to the cosmic lattice to find the particular information about you, Malcolm.

You did a very similar thing when you connected with the living piece of bamboo plant, not even a flower, that you read very successfully for Hazel, the person who had brought it to the class. That is another mechanism of psychometry, when you connect with a living being such as a plant, an animal or a person, and you follow that living being's connection to the cosmic lattice to find the required information. Your Golden Angel Mikael can help you make those connections. We would like you to make that clear in your writing; when a human makes a connection with the cosmic lattice, it is nearly always with the help of the human's Golden Angel, the oversoul or the higher self. That is what you talk about in your conscious creativity part of our first book. You say there are two ways of interacting with the cosmic lattice: one is the unconscious means of getting information from it, which is what you do from day to day as you create your reality as you travel your life path. But, when you want to create something special, such as a work of art, a piece of music or an invention, then you prepare yourself by the four steps that you enumerate in our book, and that preparation involves the Golden Angel or oversoul who helps you make the connection with the relevant part of the cosmic lattice.

When you held that piece of bamboo plant in your hand and connected with Hazel's information, your Golden Angel Mikael was helping you make that connection through the bamboo into the cosmic lattice. It would help in the future if you remembered that, because by acknowledging the guidance from your Golden Angel into the cosmic lattice, you are strengthening that connection and allowing more information to flow through it. In one of our earlier messages that we gave you that you incorporated in the book, we talk about exercising that ability of connecting with the cosmic lattice. The analogy we used at that time was that it was like a muscle and that the more often it was used the stronger it gets. We want to reiterate now that if you recognize the steps that occur when you are doing psychometry and that you recognize your Golden Angel's contribution to that, it will strengthen the connection and everything will work so much better. Also your Golden Angel will be delighted to help you do that and have the recognition of how it works. Mikael is always helping you in all these things anyway, but the mere fact of you recognizing his help greatly strengthens it. Of course we want you Malcolm, as part of your writing, to make this known to other people so that they can strengthen their connection with the cosmic lattice through their own Golden Angel. That is another chapter in the book; we've been talking about plant-assisted connections with the cosmic lattice. It's a way of finding out information about other humans.

Now we come back to regular psychometry and we say that similar strengthening effects come from the assistance of the

Golden Angel in that case. When you were tuning into the crystals in the metal pendant that Alison gave you, you were assisted in that process by Mikael your Golden Angel. Whenever you send out from your questioning spirit the electromagnetic loops of light – like insect feelers – to connect with the information in the crystals, then that process is facilitated by your Golden Angel. This applies to everyone: the stronger the connection you have with the Golden Angel the stronger will be the connection with the crystals in the object and the information contained within them. When you are talking about psychometry of non-living things, then, as in the case of the living objects, the Golden Angel assists in establishing a path into the crystals in the object so the information can be read out by the psychometric observer – which is you Malcolm or the person using the technique.

We have now an introductory chapter based on your experience and that divides off into two directions: one is where you have crystals carrying the information in a physical object and that direction includes reading out information from bones, because there you are talking about the apatite crystals in the bones which are carrying the information. That is one line of discussion and writing. Another line of writing will be the use of living organisms to make the connection with the cosmic lattice as we just discussed. Then a third line is the use of water as a medium for retention of information. Of course, you have quite a bit of information about that already. For example, we referred today to the homeopathy work, and the details of the mechanism we had given you in November last year will help you with that. You have a lot of information about Dr. Emoto's work that you can bring in. Of course, you have your own physical experiments, which showed that the water had different surface tension and/or viscosity characteristics after it had been exposed to good vibrations compared to what we will call bad vibrations, or less harmonious, shall we say, would be a better descriptor.

So there Malcolm, you have three lines of development: psychometry of solid physical objects, psychometry of living physical objects and psychometry of water. That should be enough material for our second book. These areas are well-known from the phenomena point of view but no one – that we are aware of and we are aware of pretty well everything that goes on in the human mind – has explained them in a systematic way. And the honour falls to you, Malcolm, to be able to do that as a result of the information which you have been given.

So with that, we think we have finished today, we hope this helps you plan your next book. You have a framework now on which you can plan the book, and we hope that it will flow quite easily from this information that we have given you. Of course, you can bring in all the previous transmissions of information that we have given you since you completed the first book so well. We have given you four or five other transmissions, partly because you enjoyed the process so much and we wanted to make the continuity flow from our first book to your second. We just had to make a small deviation in that course of the second book's preparation to bring in your workshop of January 22, 2011. Also, you had to bring in the experience of the psychometry workshop on January 29, 2011. So you have been through a fairly intensive time in the last two weeks, which has given you a lot more understanding and clarity of thought. In addition, you now have personal experience of psychometry, which you didn't have before, so it was well worth diverting to achieve that experience.

Once you have experienced it, a person like you, Malcolm, can constantly re-employ the same technique and do readings for people on any object. If you recognize the contribution of your Golden Angel, that will greatly increase your power and you will strengthen the "muscle" of your psychometric technique which is really connection with the cosmic lattice. You were wondering about the crystals in the metal pendant and their relationship to the cosmic lattice – we saw that question come up in your mind. Of course the crystals in objects like the pendant are connected with the cosmic lattice, it is everywhere, it is within the atoms that form the crystals. So when you go into the crystals in the metal of the pendant, you are also connecting with the cosmic lattice. You could say the cosmic lattice network is crystallized in the crystals of the metal. You are connecting with a special part of the cosmic lattice which shows up in your physical world as items you would call crystals.

Rupert Sheldrake has talked about the morphic fields, not only of plants and animals, but also of crystals. Each crystal has a morphic field attached to it, and that morphic field is part of the cosmic lattice. So everything comes down to the information on the cosmic lattice, ultimately. We would say that the physical crystals in metals are a special case in which the cosmic lattice is crystallized into physical form. There are technical differences between the material of the cosmic lattice and of the crystals. But we will not go into that now, because it is not necessary for you to talk about those distinctions. Just tell people in your chapter on reading crystals that the crystal itself holds the cosmic lattice in its structure and that will be sufficient detail. You need not concern yourself with the structure of these things between physical reality and reality of the 5th and 6th dimensions.

We think we answered your question. We are glad you raised that point about how the cosmic lattice related to the crystals in a metal. Now, we think we have given you enough to get started Malcolm. If you can discipline yourself to write for an hour every day, this work will go extremely fast. If more detail is needed, we will be happy to come and provide it. Anytime you are uncertain about one of these points just write your question on a Post-It note and put it on the door of your computer cupboard as you did with your fractals question and we will be there very quickly with a lot more information about that special point. Things will probably come up that need unraveling and explaining a little more, and we will be very happy to provide you with any information you need.

All that remains now is for us to thank you once again Malcolm for taking part in this work, for lending us your mind and body in translating this information from our dimension into your dimensions. This is great work you do to help your fellow humans and we are very grateful that you are prepared to do this work. We thank you and send you our blessings. Until we meet again much love from Crystal Light.

Elapsed time 45 minutes.

Malcolm K. Smith / 12 February, 2011

Appendix 8:
Angel Communication, April 21, 2011

Hello Dear Malcolm. Hello, hello, hello. It's good to see you back with the recorder in your hand, ready to talk for us. We are happy to be here again and we come in response to your request for help in organizing some material in your next book – Book 2 as you call it – the book about energy in physical objects. You asked if we could help you organize information on how to read out energy from crystals. You have written the chapter on energy in crystals and have come to information from our previous channelings about how to read out the energy. Also, you had information on that in our first book – how to read out energy from the cosmic lattice. So we are here today to talk about reading out energy from crystals and from the cosmic lattice.

The first point we want to make is that all of this reading is done with the help of your Golden Angel. In all cases the Golden Angel – Mikael in your case Malcolm – goes into the crystalline object, e.g. a metal pendant, and brings out the information that is required. It seems like the person doing the psychometric reading is actually reading the energy out of the crystal but in fact that process is greatly facilitated by the Golden Angel of each person doing the reading.

In your case Mikael always assists you in reading out the energy. He goes into the crystal and, as it were, decodes the symbols that are within each unit cell of the crystal and brings them back to you – telepathically of course, there's no actual traveling to and from the crystal. His intent goes into the crystal, finds the necessary information and feeds it back to you. That is why when you do psychometry and you have the words about the owner of the object, it seems as though those words come from some other person or source. That is in fact true because the words come from your Golden Angel Mikael. He is facilitating the reading process and passing the information to your spirit who relays it through your body's DNA by magnetic impulses to your brain where it becomes conscious. That is how the routine of psychometry reading works.

For anyone who is wishing to learn how to do psychometry it is preferable for them to obtain the help of their Golden Angel. Now this is where it divides into two streams:

- In the first case the person doing the psychometric reading – we will call them the reader – is aware of their Golden Angel, as you are Malcolm, and they acknowledge the fact that the Golden Angel is helping them.

- In the other case the reader does not know about the Golden Angel. All that person knows is that it <u>seems</u> the information is given to them when they focus their intent on the object which they are reading Although it is more correct to say they focus their intent on the crystals in the object.

As we have discussed earlier, in the objects there are crystals, and the information is stored in the crystals in the objects. In your case Malcolm, with the spiral metal pendant that belonged to Alison, the information was in the crystals of the metal from which the pendant was made. You were able to read out that information, just as if it were words in a book, because Mikael was translating the information from the crystals into your consciousness. That's how all psychometry works, either the reader is not aware of the Golden Angel's help or the reader is conscious of the Golden Angel's presence and help. In our opinion, in the case of the conscious awareness of the Golden Angel, the skill of psychometry is more reliable, because the intent of the Golden Angel is recognized and that strengthens the link between the Angel and the reader.

That is a summary of psychometry of physical objects and crystals, with the understanding that the physical objects all have crystals in them in some form, which you will discuss later in the book. In those crystals is where the information is stored, as we described and you have written so well about in this first chapter of Book 2.

Now we come to other skills similar to psychometry but not quite so obvious. The second skill we will talk about is that of flowers and plants being used as a link to information about the person who laid claim to the flower or plant. In the workshop you did recently, and another you did several months ago, each person was asked to bring a flower or plant. You yourself Malcolm, had a piece of bamboo in the first one that was brought by Hazel, and in the second one a form of daffodil brought by the same person, Hazel – just a little reinforcing coincidence there – we are smiling as we say that, because you understand what we mean by that. In all those cases we will call that plant psychometry, although that is not a generally accepted term. In this case, the plant is a link between the person who brought the plant, what we will call the owner of the plant, and the person doing the reading.

The information in this case is stored on the cosmic lattice and it is in a particular part that relates to the person that brings the plant. Now that plant isn't owned by the so-called owner, it is just being used as a tool to assist the plant psychometrist to make the connection. As you said in the first workshop, you only had the plant in your possession overnight, and that hardly seemed time to distill a lot of information about you to the plant. So it is not the plant itself that carries the information, but the plant provides a pathway to the part of the cosmic lattice where the information about the owner – Hazel in this case – is stored.

The reader, the plant psychometrist, is receiving the assistance of his or her Golden Angel once again, just as in the case of the object psychometry. In some cases, the plant reader is aware his Golden Angel is making the connection, and in some cases, the reader is not aware of the connection. But we assure you that in both kinds of case, the connection is always made by the Golden Angel. It is more necessary in this case that the Golden Angel helps make the connection, because the link between the plant reader and the plant owner is more tenuous – it is not such a direct link. It is not a case of going into the plant to find the information, it's a case of the Golden Angel using the plant as a pathway or connecting link into the cosmic lattice. That is the way it happens. The plant reader sits quietly with the plant in hand. The plant acts like a lens that focuses the attention on its owner. The Golden Angel of the reader follows that focus into the relevant part of the cosmic lattice, mines information about the owner on the cosmic lattice and brings it back to the reader. That's why, Malcolm, in the case of Hazel, you received information which was quite accurate, such as Hazel was very interested in aromatherapy. That particular bit of information was held in Hazel's part of the cosmic lattice – and a lot of other information about her too – and your Golden Angel brought you some of that information as proof that you were able to use the plant to do a psychometric reading. This technique is quite distinct from object psychometry, where the Golden Angel goes into the crystals in the object.

There is yet a third technique which you have experienced – and a fourth which we will talk about – these two techniques, which we have not discussed yet so far, are based on a similar process. The first technique is that of coloured ribbons, and in that case the relative colours of three ribbons that are chosen from a set of seven is enough to set up a pathway to the part of the cosmic lattice concerned with the person that chose those three colours. Once again, the Golden Angel, either recognized or not by the reader, goes into the part of the cosmic lattice relevant to the person that set up the three ribbons, and brings back information that is relayed through the psychometrist, who gives a reading on the "meaning" of the colours of the three ribbons in relation to the characteristics of the person that's arranged the three ribbons. To say it again, you have the arranger of the ribbons and the reader of the ribbons, and the Golden Angel is the link between the arranger's part of the cosmic lattice and the reader.

The other technique that you experienced once is reading an impression in sand of someone's hand. The technique comprises a tray of sand into which the person who wishes to have a reading puts a hand and creates an impression. Then the psychometric reader senses, from the impression in the sand, information about the person who made the impression. This works just like the coloured ribbons, in that the Golden Angel of the reader goes into the section of the cosmic lattice that contains information about the person who made the impression, and brings that information back to the person who reads the impression. So this technique has a very similar mechanism to that of the coloured ribbons.

Those are all the techniques that you wish to know about. There are some others which are variations on the same theme that we have described here this morning. We will not go into those other techniques now and confuse the issue. They all work on the same basic technique, which is connecting with information held either in the object of psychometry or in the cosmic lattice – we should say the part of the cosmic lattice relevant to the person requesting the reading, which is usually the owner of the object or the plant, the arranger of the ribbons or the maker of the impression in the sand.

The reason it works this way is that information about each person in the world is recorded in their own personal part of the cosmic lattice. That part of the cosmic lattice is connected with the person's multidimensional DNA in dimensions other than your four space-time dimensions. There is a strong connection between the person's characteristics, their own personal DNA expressed in the dimensions 5-12 and the cosmic lattice. The cosmic lattice is in all 12 dimensions of the universe and the person's DNA is expressed in all 12 dimensions, so that when a Golden Angel goes to the relevant part of the cosmic lattice for information about a particular person, a connection is made with that person's DNA. That is why it is so important to have an understanding of the multidimensional character of each person's DNA, that is why you are learning about this from Kryon's book about the multidimensional characteristics of DNA.

These two things are linked, the cosmic lattice and each person's DNA. Each psychometric reader's Golden Angel has access to the cosmic lattice, and through the cosmic lattice to the multidimensional DNA of the person who is being given the reading – usually the owner of the plant, the arranger of the ribbons or the maker of the impression in the sand. All of those peoples' characteristics are recorded in their own personal DNA, of course, and the Golden Angel can get access to that through the cosmic lattice. That's how those latter three techniques – plant psychometry, coloured ribbon arranging and sand impression – all work. They all work through that link which the Golden Angels are able to make between the cosmic lattice and the personal DNA of the person who is seeking the reading.

Now in the case of psychometry of physical objects, the information is actually in the object. The information is stored in crystals in the object, and has been put into the crystals by the cosmic lattice. The cosmic lattice acts as a conveyor belt, if you will, conveying the information into the crystals as each event occurs. Let us take as an example Alison's pendant. You, Malcolm, reading the pendant psychometrically, saw Alison walking in England, and you knew it was England, because you saw a south coast promenade and pier. That event was placed in the crystals of the metal of the pendant by the cosmic lattice, which acts as a conveyor belt and brings events into the object. The cosmic lattice makes it possible for the information to be laid in the object's crystals. When the Golden Angel goes into the object's crystals, it is via the cosmic lattice. Since the cosmic lattice puts the information into the crystals, it is the obvious route for reading out the information.

The basic difference between psychometry of objects and the other three techniques we talked about is where the information is stored. We can summarize by saying in all cases the Golden Angel of the psychometric reader is the main researcher, or the active principle, that goes and gets the information from the place where it is stored. The place where the information is stored is in two different locations:

- In the case of physical objects, like pendants or jewelry, the physical storage is in the crystals within the structure of the object – it is understood that the object may be metal or bone, or even human flesh, we will discuss those more subtle differences in the way the material is organized in later chapters. In any case, it's within a crystal in the structure of the material being examined.

- The other case involves the techniques which use flowers, plants, arranging colours or making impressions in sand. Then, the information is stored in the subject's personal DNA – the multidimensional part – and in those cases, the Golden Angel, acting as messenger on behalf of the reader, gains access to the subject's DNA through the cosmic lattice.

The cosmic lattice is the common denominator. It puts the information into the crystals, in the case of objects, or it's the link with the multidimensional DNA of the subject individual. We can see that you are thinking of a diagram here, in which two storage places are shown, either crystals in an object or the multidimensional DNA of a person. In both cases, the cosmic lattice links to those storage places and provides a path for the Golden Angel to get to the information.

So Malcolm, we think we have described the situation, which is relatively simple when it's broken down like that. We realize it's confusing for someone trying to understand the physical part of the technique – it seems that there is no connection between an object and an impression made in sand, so how can all these things lead to similar results? Now you understand the different arrangements in what is really a quite simple set-up. So now you can explain that in our book.

We hope this has provided you with clarity and you are able to transcribe it into a form that you can use as a reference as you have in our first book. This will help you complete Chapter 1 of Book 2.

Once again, we thank you for helping us present this information to humans. We are very pleased that when you don't understand something with the clarity that you would like to have, you come and ask your Guides for help and they are able to connect with us and set up a session like the one we have had this morning. This is an excellent way of clarifying issues and making it all quite plainly understood.

We thank you Malcolm for helping bring our words to humans. You give intent every day to help your fellow humans and you are helping them greatly by making this information known in the physical world. We look forward to helping you write all this out. If there are any parts you do not understand, then go on the pendulum again and we and your Guides will help clarify any points that are not clear. But we think you have the essentials. We thank you for doing this work, Malcolm. As always, we are very happy that you are willing, interested and fascinated by this work. We know you love doing it and you give thanks for this work. We encourage you to push on and finish Chapter 1 and get it edited by Dawn.

Now we leave you with all our thanks and any blessings that are ours to give you and with all our love. Goodbye.

Elapsed time 40 minutes

Malcolm K. Smith / May 27, 2011

Appendix 9:
Angel Communication, July 7, 2011

Thank you Malcolm, thank you for agreeing to do this at short notice. We have seen that you have been studying, studying, studying this work mainly from Kryon about the cosmic lattice, the crystalline grid and the crystalline sheath, and there is some confusion in your mind about the precise differences and relationships between these different items of the universe's structure. So we have come here today at your request. We've been expecting this request for some time. We made sure you had batteries and tapes ready to be pressed into service this afternoon when you were alone so you will not be disturbed as you make this recording. We will start now to explain all these terms.

First, we will start with the cosmic lattice, which you are most familiar with. We think you have a good description of that from your previous talks in which you refer back to Kryon. Plus, you have a very precise description in our first book. Not much in addition to that is needed, except we will say that the cosmic lattice is a construct of energy throughout the universe. It consists of 12-sided honeycomb cells with each side of each cell in a different dimension. This is a source of great energy, although it doesn't appear to be so, because the energy is balanced against itself. This energy could be released for the service of humans just as it is used by many alien races in the universe. Humans have not developed to the extent that they can use this energy yet, but a time will come when they can use this free source of energy that is infinitely replaceable.

In addition to being a source of energy, the cosmic lattice is a means of communication between the far ends of the universe that is almost instantaneous. Unlike light, which takes years to travel astronomical distances, a disturbance of the cosmic lattice travels instantly from one end of the universe to the other. This is the basis of telepathy between humans, between animals, and between humans and animals, which all takes place by this impulsive energy which can travel from one part of the universe to another apparently instantaneously. When someone that wants to communicate with you needs to get your attention, they can do so by means of the cosmic lattice. You suddenly become aware of a voice in your head or a feeling that something is happening. You have had lots of instances in which you experienced this sort of thing, Malcolm. We note one in passing which was very strong: you and Beth were driving from Deep Cove in separate cars when suddenly you, Malcolm, received a message that Beth needed to do some shopping in SuperStore. So you drove into SuperStore parking, trusting your intuition, but then you saw Beth drive by because she did not think that you would do that. When you saw her later, she confirmed that she did want to stop, but thought it was too late to get a message to you. So you know the strong feeling that comes from messages on the cosmic lattice.

This is the way that animals react to things that are happening in the world around them. For example, the elephants and other creatures in Sri Lanka that responded to the earthquake-generated tsunami which hit the island on Boxing Day, 2004. The animals anticipated the tsunami and went to higher ground. Unfortunately, humans did not trust their intuition, although they received the same signal as the animals, and many were killed.

Also in the cosmic lattice is a record of what all humans have done and plan to do (see *Spiritual Chemistry* by Malcolm K. Smith, page 97). The events of each human's lives, past and future are recorded there. – We are referring here to what humans call past and future lives. You know, Malcolm, that there is no such thing as past and future lives, but this is how it appears to humans, and so we honour that perception, because it makes it understandable for them. In addition to the deeds, wishes and plans for human lives on Earth, there are what Rupert Sheldrake (see his book, *Presence of the Past*) calls morphic fields which hold "plans" for making all the species of animals, plants and crystals that are on the earth. All these things are accessible to humans.

Humans intuit information from the cosmic lattice every second of every day in the unconscious creation of their reality. In addition to that, you have the conscious use of the cosmic lattice in creation of something new, like an invention or a work of art. For example, Malcolm, when you want to create something, you get the help of your Golden Angel Mikael to go into the cosmic lattice and bring back to you the information required for your creative act. That information you perceive as inspiration coming from outside of you. So you are quite familiar with the cosmic lattice, Malcolm, and we will leave it there and move on to the crystalline grid.

This is the network of magnetic lines that are around the Earth, and it is part of the cosmic lattice. We will emphasize that part again; the crystalline grid is part of the cosmic lattice. The crystalline grid around the Earth is a subset of the cosmic lattice, because whereas the cosmic lattice spreads throughout the universe, the crystalline grid is concentrated around the Earth and its environs in space. The crystalline grid doesn't extend all the way to the other end of the universe, it is localized for the use of the Earth, although it is part of the cosmic lattice, as is the crystalline sheath that we will be talking about as well. These things are part of the cosmic lattice, which is by far the overall grand network of the universe. The crystalline grid and the crystalline sheath are subsets of it. The crystalline grid is the subset that is around the Earth, and the crystalline sheath is the subset that is around the DNA molecules in every cell. The purpose of the crystalline grid is to facilitate humans on the Earth. It is the human race's part of the cosmic lattice and is there for their purposes during their lifetimes on Earth. It has been referred to in some situations as the akashic, or the akashic record, which are just other names for the crystalline grid. The akashic tends to talk about the past record of human events, all the things done by any human in any lifetime lived on the Earth.

We could say that the crystalline grid is humanity's special corner of the cosmic lattice. This is the part that is connected with the crystal skulls. We know, Malcolm, you developed your interest in them when you read the book about the skulls (See: *The Mystery of the Crystal Skulls* by Chris Morton and Ceri Louise Thomas). It was apparent to you that there was a connection with the crystals you were talking about in Chapter 1 of your new book and the energy that is stored in them is paralleled by the energy stored in the crystal skulls. You know now, as a result of reading Kryon, that the crystalline grid is anchored to the physical body of the Earth by the crystal skulls. That was put in place at the time of Lemuria and later modified in the time of Atlantis. Those crystal skulls were made by matter creation in the shape of the human skull. Just as Sai Baba (See: *Miracles* by D. Scott Rogo) can materialize physical objects in the modern world, so the Lemurians and some of the Atlanteans could materialize objects and that is how those crystal skulls came into being. The purpose of them is to anchor the crystalline grid to the surface of the Earth and act as a "road map" for humanity's development (See: *Kryon Book 7 – Letters from Home* by Lee Carroll). Also they record past events in the very long life of humanity on the Earth which goes back 100,000 years (See: *Kryon Book 12 – The Twelve Layers of DNA* by Lee Carroll), to the time when people from the Pleiades came and interbred with humans and changed the human multidimensional DNA to include "the search for God."

One more thing that we should add is that whereas the crystal skulls hold the past record of human events, and future plans, the day to day changes in the progress of humanity – which is creating its reality as it goes – are signaled to the crystalline grid by the sea mammals, the dolphins and whales. They are the other anchor as related by Kryon.

So that is a description of the crystalline grid. Now a further subset of that, which is not only very personal, but is concerned with every cell of your being – and there are trillions in every human body – and each of those cells has a crystalline sheath around the DNA molecules. This word, "crystalline," in the case of the crystalline grid and sheath, is not a crystal in the same sense as the physical crystals that can store energy you were talking about in Chapter 1. The crystalline grid and crystalline sheath consist of information that is crystallized and of course information is energy. Do you see the difference? Physical crystals exist purely as crystals, and into those energy can be placed, and stored, and read out and removed. In the case of the crystalline grid and sheath, information is lined up in an orderly way so it appears like a crystal, which is why these items are referred to as crystalline. The result is very similar but the origins are different.

The crystalline sheath is around each pair of DNA molecules. The DNA consists of the two physical strands, carrying the genetic code, and ten magnetic strands – not recognized by Earth science yet – which are coiled together in a convoluted way (see: *Kryon Book 6 – Partnering with God* by Lee Carroll). Around all that is the crystalline sheath, which carries information specific to that particular human. Just as the crystalline grid has information specific to the Earth, so the crystalline sheath has information specific to a human, in particular, to a specific cell in that human's body. Although there are slight differences between sheaths around cells in different parts of the human body, essentially all the cellular crystalline sheaths have similar information. In each sheath is recorded information about that human's past lives, the plans for this life and future lives – which we refer to as karma – and a lot of information about the state of the human's DNA at the present time and how it changes as that person ascends or approaches graduation, another word for ascension. All that information is collected together in the crystalline sheaths around all the DNA molecules in all the trillions of cells of each human body.

So the crystalline sheath is a subset of the crystalline grid which is a subset of the cosmic lattice. They fit inside each other, as it were, from an information point of view – a bit like Russian dolls fitting inside each other. That is a good analogy for remembering the relationship between these three items, they nest inside each other: the crystalline sheath and its information is nested inside the crystalline grid which in turn is nested within the cosmic lattice. That picture, which we know appeals to you, can be used in our book to describe how these things all fit together. The common denominator in all this is the concept of "crystalline" in the sense of ordered structure.

What Kryon is talking about in the book you were reading last night (see: *Kryon Book 7 – Letters from Home* by Lee Carroll) is the way that magnetic signals come from the cosmic lattice, and other parts of the universe via the cosmic lattice, into the crystalline grid for general dissemination among all humans. If there is information for one specific human, then it comes into the relevant crystalline sheath, and that information in the form of magnetic pulses and waves is transferred to the magnetic strands of the human DNA. The magnetic strands induce an electric current in the physical DNA, which is transmitted through the body to the brain, which of course is monitored by the spirit, and in that process, the information becomes conscious to the human.

For example, let us consider two twins. One stayed on the Earth and the other went in a spaceship to Mars. If the Mars twin wanted to send a message to the Earth twin, he would beam that out from his crystalline sheath into the cosmic lattice, it would travel via the cosmic lattice through the crystalline grid into the crystalline sheath of the Earth twin. Then, by the process of electromagnetic induction it would become an electric signal in the body of the Earth twin, where it would reach his consciousness. We recognize that as the process English-speaking humans call telepathy.

It is a similar mechanism that the Golden Angel of any particular human uses when the human wants information to create something new – say an invention. The human requests the Golden Angel – this is usually done unconsciously, although the Golden Angel is connected into the "circuit" by the work put into the creation – to go to the cosmic lattice and find information on the thing of the invention. The Golden Angel comes back from the cosmic lattice with the information and transmits it into the crystalline sheath around the DNA in each of the human's cells, from where it comes into his consciousness. The human perceives that information as something coming from outside his body, and he says he is inspired with the information about his creation. In fact, it is partly the human's work and partly the work of the Golden Angel. But since all humans are aspects of Golden Angels, this process works smoothly and beautifully and there is no question of ownership because a cooperative effort is bringing the information from the cosmic lattice into human knowledge.

The mechanism of crystal psychometry is quite clear now. You know that the Golden Angel goes into the physical crystal itself and brings back the required information. You already have in our book the example of how when you held Alison's metal pendant, from the crystals in the metal you picked up the scene of Alison walking on a promenade in an English seaside town – you saw the pier. That information was brought to you by Mikael your Golden Angel, although you did not realize it at the time, but you do now of course. That is the basic process of psychometry.

Additional techniques consist of: flower psychometry, coloured ribbons psychometry and sand impressions psychometry. In all those cases, the Golden Angel of the reader goes into the part of the cosmic lattice where he finds, for example, the crystalline sheath of the owner of the plant.

(Malcolm, just to make that clear, you were reading a plant which was brought to the sensitivity class by Hazel. So your Golden Angel Mikael went into the part of the cosmic lattice that contains the crystalline sheath around Hazel's DNA. From there, Mikael gave you some information about Hazel, for example, that she was interested in aromatherapy. The information was brought back by Mikael from Hazel's crystalline sheath, through the cosmic lattice to your crystalline sheath, and that information was made conscious in your body. So you were able to write down, "interested in aromatherapy" as a characteristic of Hazel. When you gave that item of information to the class – at that point you did not know who had brought the plant – Hazel confirmed that she was the owner of the plant and that the information was correct.)

This is the way the process works. All these crystalline structures are just energy holders. They can be physical, as in the case of a quartz crystal, or they can be "information crystals," in which the information is lined up in an orderly way.

Now, we think we have covered all the relationships that you wanted to know about, Malcolm, and that you have enough information to explain the relevance of the crystal skulls in the first chapter of your new book. You can refer to the nesting of the crystalline sheath within the crystalline grid within the cosmic lattice, and how the information is transferred back and forth. You can refer to later chapters to talk about plant psychometry, for example. We suggest that you cover the basic structure of the sheath, grid and lattice in the first chapter, and give the refinements about the Golden Angel using the crystalline sheath of other people for a later chapter, because it may confuse some people at first. In your first chapter, just talk about the nesting structure and the relevance of the crystal skulls to that, and how the Golden Angel goes into the physical crystal of, say, a metal pendant, and brings back the information which the reader perceives as coming from outside his body and therefore feels inspired.

With that, we leave you today. We thank you for asking for this clarification and we are very pleased to be able to present it to you and clear up all those questions you had. Thank you for mentioning this to your Guides, who were able to connect with us and bring about this little talk we have had this afternoon. As always, Malcolm, we are very happy to provide the information you need. At any time we are pleased to come and talk with you and to bring this information through you to humans in the form of our books. Don't forget what we told you earlier this year, that these books will be very necessary in the new Earth, and you are able to clarify these ideas very well here in the old Earth because of your learning activities here in schoolhouse Earth.

Now we leave you. We thank you for all your work and trying to understand these structures which appear convoluted to a human. When you are in the 5th and 6th dimensions, as we are, these things are quite clear. So, we can bring this information and with the help of analogies – like the nesting Russian dolls – we can explain these things to you, Malcolm, and you can pass on those explanations to other humans through our books. You give a valuable service conveying this information to your fellow humans and we thank you for this work you do. We are delighted by the fact that you continue to write and show great interest in this knowledge that is available to any human if they just ask. We are so pleased that you ask Malcolm. We thank you and send you our blessing and our love. Goodbye from Crystal Light.

Elapsed time 50 minutes.

Malcolm K. Smith / July 19, 2011

Appendix 10:
Angel Communication, October 25, 2011

Hello Malcolm, once again we come together. It's been quite a long time since we last did this with you and we are very happy to be back here working with you again on this project. You have had some difficulties finishing the chapter on the pilgrims travelling to Santiago de Compostela to view the bones of Saint James, and we are here to help you with that.

But first we wanted to give you some general information that was started this morning by your Guides. When you were born, you were earmarked to do this work. You had special circumstances set around your birth in this lifetime. Just as your Guides told you this morning, you were marked for a very long life with good health and high energy. A long life as an example to other people that even when you are over 100 years old, you are able to run a very active life and do very useful work, even though most people expect to be dead at that time. All this is possible if you believe in Spirit and you actively work for Spirit, as you have done, Malcolm. When you were about 12 years old you were told not to smoke and to take up distance running; you did both faithfully and that has given you a very healthy body. The few diseases that you have had, or weaknesses that have occurred in the last year or so have just been in preparation for your continued life under the new energy conditions that are coming to the Earth.

You see by the political upheavals – we call them political, people searching for democratic rule – and the physical changes that are occurring, like earthquakes – just recently, there was a bad one in Turkey. And the weather changes, with storms and floods in some parts of the world, and fires in other parts, all these things are indicators of a time on Earth when special energy is coming to the Earth. To take advantage of this energy, you needed to have these medical experiences which would strengthen your body and your physical mind and prepare you to usher in the new energy, which is largely exposed to the fifth dimension.

Next year, 2012, you are going to see the thinning of the veil between the fourth and fifth dimensions, and this will bring about a blessed energy that people who are prepared will be able to take advantage of. You are one of the people that are being prepared; there are thousands around the world like you who are experiencing similar things at the present time. That is why Alannah wrote to you about people who are having symptoms of Parkinson's disease. They don't really have Parkinson's disease, just symptoms of it, which are part of the preparations for them to take advantage of that new energy that's available to them.

You are like those people, you are preparing for your post-retirement lifetime. You know in your case, you had a rebirth in 2002 (May 20 – your mother's birthday), when that occurred, it set you up ready for a second lifetime in the same body. What you are experiencing now – the apparent Parkinson's disease and the other smaller medical symptoms – are all part of the preparations for growing from the adolescent stage, which you are in now, as you learn to use the new energy. Next year, when the new energy is poured into the world, you will be able to take advantage of it because then you will be an adult for the second time in the same body.

What your Guides were telling you this morning was all part of the same story, that you are protected from certain diseases like cancer, that would bring your life to an early close if you were to catch them. In your case, you are protected from cancer and you are assured a healthy life all the time you continue to watch your diet, exercise regularly and – this is the most important part – all the time keep a strong contact with Spirit, which you do. If you follow those few rules, including of course no smoking and keeping up the running, if you follow the rules as we set them out, then you will be assured a lifetime well in excess of a hundred years. That East Indian man that ran the marathon in Toronto recently is an early example of that. He's doing his part to let people like you, Malcolm, know that it's possible to be in advanced years and yet still have great energy, great endurance and great strength, because your body is fired with the love of Spirit, as his is. That love of Spirit makes incredible things possible. So continue with your good practices of running regularly, watch your diet, don't smoke, drink only moderately, and your body will serve you well for many, many years to come. You will be a fine example and a help to your fellow humans.

That was all we wanted to tell you about the special circumstances of your second birth and how things were set up before your first birth, before you came to Earth so that you were prepared to accept these changes at this time of your life. You are 75 years old at your next birthday in November and there will be a great celebration in life for you. Many things that have been promised will be poured into you at that time and you will be ready to lead an exemplary life showing other people the love of Spirit. Part of the duties you have agreed to is in writing a number of books for us. The first book *Spiritual Chemistry* that you wrote was the first of those that you agreed to before you came here. Certainly while you were a young man you agreed to write those books and you delivered on your promise to write that first book to get started. Now you have a process ready at your finger tips for writing a number of books and you are well into our second book. That's what we want to spend the rest of our time with you talking about.

The chapter you have nearly finished is about energy in crystals of apatite that are in bones. In the case of the saints, that energy is so strong and so blessed that it preserves the whole body. Even after death, the body continues in good condition because of the love energy that is stored in the apatite crystals in the bones. People that wanted to experience that energy came to places like Santiago de Compostela to be near the bones of a particular saint like St. James because they could experience that energy by being close to the bones, although they could not touch them, but just being within a few centimetres of the bones was enough for them to feel that blessed energy. That was the reason for their pilgrimage.

As we told you, people in modern times don't realize the strength of the energy that is available to them. They concentrate on the journey as the main part of the pilgrimage; they feel that the preparation and actually executing the journey is the thing that is leading them to new spiritual discoveries – and it is. They benefit greatly from that; but they could also benefit from the realization of the love energy that is in the bones at the final destination of their pilgrimage as well. They can benefit greatly by experiencing the love energy from the bones of Saint James or whichever saint they are undertaking their pilgrimage to. Malcolm, what we want you to do is tell your personal story about the feelings you had when you stood close to the bones of Saint James in the cathedral of Santiago de Compostela. You were in that tiny passage that runs underneath the altar where the bones of Saint James are stored and you experienced strong emotions that you did not understand.

That very same night, you woke up in the middle of the night in the hotel, probably to go to the bathroom. When you came back you noticed the star constellation called Orion was very neatly framed in the skylight of the top floor hotel room that you were given especially so that experience was possible. (I just learned that now – MKS.) That had a two-fold purpose: one was that you were considering Orion as the name of your consulting company at that time so it had special meaning for you. The other was that the constellation was completely framed by the window and that was an example for you of the way in which the energy is stored in the bones – it is held within a frame like the window frame. The frames in the bones are the unit cells of the apatite crystals. There is an analogy here; the frame of the window was analogous to the unit cell in the bones' crystals.

Just as the information is stored in the unit cell acting as a frame, so the constellation of Orion – which is a very large amount of information – was stored that night in the frame of the hotel room skylight. We know that the star constellation moved on as the Earth rotated and it was framed for a small fraction of time. So you see the synchronicity that as you came back to bed and you looked out of that skylight, you saw the constellation perfectly framed, that was a coincidence that was a special message to you at that time. But you didn't appreciate the full significance at that time – although you knew it was significant – until you wrote in the second chapter of our second book about the energy stored in the crystals in the bones of Santiago which you had been near earlier that day. But now in retrospect, you can see what a beautiful demonstration that was of the love that is present in the messages in the bones of saints and many special people.

That was a further example of the information that can be stored in any crystals in any material, such as the pendant that you got from Alison in the psychometry class you attended. You have learned about crystals and their unit cells holding energy, and this was a further example – this was the grandfather of examples of energy being stored in a framework to help you understand the miracle that is at the basis of psychometry. You have explained it well in Chapter 1 and we will continue to help you. We have already given you material for Chapters 3-6 and you will develop more of that as we have more of these sessions where we talk about that. Just to finish off Chapter 2 now, we would like you to tell your personal story as we just enlarged on it, and generalize on the way synchronicity can act to reveal a deep and significant secret. That is what we want you to do.

Some words that may help you are as follows: This experience of the framing of the constellation of Orion in the skylight of the hotel room is an illustration of a very deep truth that became apparent as a result of the synchronicity. (I was aware as this event happened that it was synchronous and significant for my future but I could not get beyond that point of realization at the time – I remember being very quiet and thoughtful as the bus left Santiago de Compostela the next morning but I didn't think I would have to wait 15 years (1996-2011) for the significance to dawn on me – MKS) That synchronicity was given to Malcolm as an example of the way information energy can be stored in a frame work. The energy of the Orion constellation was stored momentarily in the frame of the skylight as an illustration of the way that information energy of love and the life experiences of the saints is stored in the unit cells of the apatite crystals within their bones. So this love energy is stored like a battery – a battery that keeps providing love energy to those that come near. It's not necessary to make a precise physical connection to feel the benefit of that love energy. Instead, all that is necessary is to have faith and stand very near the bones, and that love energy will flow from the cells – it will be read out from the crystals in the bones – and be experienced by the pilgrims. Yet it does not diminish – the love energy that flows out to the pilgrims that approach close to the bones – that energy is not diminished by the reading out because it is like a library of energy. Just as you go into a library to read information from a book, you do not take the information energy out of the book and you do not need to take the book out of the library to get the information. That is analogous to the situation of how you can read out the love energy that is stored in the bones and receive the benefits of that love energy. Yet at the same time, it is not diminished and continues in great strength for other pilgrims who can come and do the same thing time after time. This is one of the great gifts of God – All That Is who provides this possibility of humans receiving love energy to their great benefit, joy and strengthening of their links with Spirit.

There Malcolm, we've given you some words that you can use to finish up the chapter. Now we have completed what we intended to do today. We thank you for arranging to do this and for looking forward to this talk with us, we were certainly looking forward to it as well. It's a very joyful occasion when we can connect with you like this and we know you enjoy it as well. We look forward to the great strengthening that comes from sharing energy with us. We are happy to have explained how your life will be prolonged. You need not worry, although you have to follow the general rules of living and you shouldn't do any dangerous things, of course. But you understand that. All the time you live a moderate life you will be held in good stead by Angels who are supporting and helping you. You will be a great example to other people of the strength and benefits that come from Spirit if you just believe.

So now we finish our session with you, Malcolm. We finish this session, but we promise you there will be many more. Many gifts will be showered upon you at your 75th birthday. You will have a happy time for many, many years to come doing meaningful work helping your fellow humans. That is all from us now. We thank you Malcolm for doing this, for making this possible, setting up your recorder and making sure everything was working well. We leave you now with all the blessings that are ours to give, with great gratitude for this time we have had together and the information we were able to pass to you.

We thank you, we bless you, we love you. Goodbye.

Elapsed time 35 minutes.

Malcolm K. Smith / November 10, 2011

Appendix 11:
Angel Communication, January 3, 2012

Thank you Malcolm, thank you, thank you. As always, you are ready to hear words from us. We are here today to tell you more about our book and the chapter you are writing at present which you call number 4 and which is about storage of information energy in cells in your bodies and in other cells as well, such as animal and plant cells, but mostly in the human body. As you have read in textbooks, there is a row of molecules of surfactant – surface active agents, generally called lipids – that form the outer wall of all the cells in human bodies. These cells have got a double layer of surfactant-like molecules (you know the kind we mean with long tails and polar heads), which is the basis of the cell wall. In the book by Bruce Lipton, *Biology of Belief* (Ref.4-2), all this is explained very well and you can use that as the basis of your description in starting to write Chapter 4.

We want to tell you today how that double lipid layer is used to contain information. You have seen it is very small, but of course it has to be small because the cells are small. The trillions of cells that make up your bodies contain the double layer of lipid molecules which is used as a chemical defence against the outside world – as it is to them – the environment in which the cells find themselves. They contain within their wall all the things necessary for life in the cell, in particular, the nucleus where the DNA is stored. These cells have a wall that goes around them in all three dimensions and in that wall is the double layer of lipids. In that double layer is what humans have called liquid crystal, because it is partly in a liquid state, it is relatively mobile so that the cell can change its shape. But it is nevertheless a crystal, because the molecules are aligned. This alignment of the molecules results in crystal unit cells which are quite flexible but usually maintain a shape which approximates a cubic type of crystal unit cell.

*(Dear reader please be careful here. Science uses the word "cell" in two different ways: 1. A unit of living matter. 2. The smallest unit of a **crystal's** structure formed by atoms or molecules – see Figure 1-2. I have called the latter a "crystal unit cell.")* But the shape is not strictly adhered to because the cell needs to move and the liquid crystal needs to flow a little to accommodate the cell wall movement, so the shape of the crystal unit cell changes slightly. Nevertheless it is still an efficient holder of information energy. The information energy is stored in the liquid crystal just as in a regular solid crystal.

We can see that you are concerned how that can be perpetuated when the cells keep changing. You are right in thinking this, but once a cell has acquired a unit of information – a byte you could say in modern computing terms – that byte becomes part of the cellular structure, part of the double lipid layer in the cellular wall, and that information is duplicated in any daughter cells that come from the parent cell. There are instructions in the DNA that tell the cell to duplicate any information that is held in the double lipid layer. This is something that is hard to explain in physical terms. Let us say that the parent cell has intent as a result of the information that is stored in its cellular wall structure. That intent is held by the DNA, so that when the cell replicates itself before its death, it does so not only replicating the physical layout of the cell wall structure, but also the information that is there because the intent is held by the DNA in the cell nucleus. In addition, there are other parts of the cell that have information stored in them, you have read of the mitochondria.

This system works well and transfers from generation to generation not only the basic information of the cell but also its acquired information. In this respect it is quite "Lamarckian" – as you know Lamarck was a Russian researcher who said that cells pass on from generation to generation acquired characteristics. This is a prime example of Mr. Lamarck's vision in seeing that cells would pass on information that had been acquired during the lifetime of the cell. Here we have a way of acquiring information and storing it in the liquid crystalline structure of the cell wall, and we can be sure that the information will be duplicated as each generation of cells comes forward, because the acquired information is passed from each generation of cells to the next cell generation. This is a beautiful system that works well to acquire, retain and pass on acquired knowledge and habits of any organism.

You have heard about organ transplants, in particular heart transplants, that pass on to the organ recipient life characteristics of the donor. For example, in *Biology of Belief* is recounted the story of a health-conscious young woman who developed a taste for beer, chicken nuggets and motorcycles after she received a heart-lung transplant. Her research showed she had received the heart of an eighteen-year-old motorcycle enthusiast who loved chicken nuggets and beer. That was simply because of the information that was stored in the heart cells, in the crystal structure of the heart cell walls. This information in the donor's cells was passed on to the woman now that she was using the heart that he had grown. Not only the heart, but all his life habits and preferences were passed on as well in the walls of the heart cells. So this is a very powerful memory tool that can provide surprising information.

It turns out that these organ donor stories that you have read are a real basis for information transferal between generations of cells in the body. It's a part of being the multi-cellular organism that humans are. Every cell in the human body has to die and be replaced, and if there were no way of passing on the information acquired by one generation of cells, then there would be no learning for the whole organism. So it is necessary to have this process by which not only the structure of the cells is passed on and replicated, but also the acquired information – if you like the memories of the cells – are also passed on through the DNA memory retention system which is quite complex. As you know there are twelve layers of DNA, and in some of those layers is the mechanism which ensures that the cell memories are passed on from generation to generation of the cells. This is a beautiful mechanism that works without any trouble at all. But because human scientists are so certain that memories are associated with the brain, they do not see that every cell in the body of a human – or animal – has its own experience and memories and passes them on to subsequent generations of cells. It is very necessary that this should happen because some cells live in the body for only a few days, or a few hours in some cases. So it's very necessary that the cells not only have a memory – a cellular memory as Kryon calls it – but also there is a mechanism for passing on the memories, so the new generation of cells learn from the experience of the older generation of cells.

It's a bit like humans. One generation of humans adds to the education system of the upcoming generation. You yourself, Malcolm, teach young humans about science. You pass on the information that you have acquired about science as a result of your life as a researcher. You pass that on to the next generation – maybe the second generation since you are the grandfather now. You pass on the information to the generation that is taking over the world, and in the same way each cellular generation passes its information on to the next generation of cells that is to form the bodies of humans. It's a learning process that never stops because each generation receives the benefit of the previous generations' experiences. That's why Mr. Lamarck was quite correct in his beliefs that acquired characteristics were inherited, although a lot of people doubted him at the time because they could not see a mechanism. More and more scientists, biologists in particular, are now seeing the wisdom of Lamarck's perception that this must be so to explain how the specialized behavior of certain species can be developed. It's no good learning by experience if the learning of one generation is not passed on to the next generation, whether it be humans or cells. It is the same process working in each case.

When you are writing Chapter 4, Malcolm, we need you to start off with this discussion of cells passing on their acquired memories and characteristics to the new generation of cells. Not only does a heart cell need to know about how to operate as a heart cell – the contractions of a heart go on for years and years – but also the heart cells need to acquire a knowledge of the whole organism's experience and to hold that in their cellular memory. They pass on that knowledge to the next generation of heart cells, which benefit from the experience so that acquired characteristics are passed on. That is why if a heart is taken out of a donor body and put into a recipient body, it still carries memories of the donor heart cells, which get passed on to the recipient. Sometimes the recipient acquires surprising habits or characteristics, as in the case of the beer and motorcycle lady we told you about earlier. But it all makes good sense that it should be like this, and this is a mechanism that has been created by God and it works very well without any interference from other sources, because the acquired characteristics are locked away in the cellular memories. Even if a section of those cells is lost, the memories from the cells transfer to the DNA and pass from human generation to generation through the transfer of DNA that occurs at the conception of a child.

You know from Kryon's description that the cellular memories are passed down from human generation to generation. In addition, cellular memories make it possible for animals and birds to acquire such amazing specific memory patterns such as bird calls and habits like feeding on platforms containing food that humans have put out. The latter habits have been acquired in recent times, there were no situations like this before. Some birds acquire special habits; we are reminded of the blue tits, those little birds in England that break into aluminum foil caps on bottles of milk left on door steps so they can drink from the milk. The ability to make a hole in aluminum caps to get a drink of milk is an acquired characteristic that is passed on to the new generation. As soon as a bird is hatched from its egg it knows it can get that kind of food in that particular way. How do you think that knowledge is transferred from one generation of blue tits to the next? It's through cellular memories such as we have been describing.

It's a very powerful learning tool for people, because the human race would not have learned so quickly if it had to depend on teaching. Habits of behavior that it had acquired over generations through teaching from one generation are important of course, because the brain is involved in that activity. But at the same time, there are acquired characteristics that are not taught, but which are passed on from generation to generation. Music is a case in point here. We know some music is taught in schools, but many races have an acquired love and knowledge of music that is passed on through the memories of the cells in their bodies. So you could say that music is in their blood and that would be very true – literally.

This is how we want you to start off Chapter 4. First of all, discuss the cell wall and the double layer of lipid molecules and the flexible crystal unit cells that are in those liquid crystals. Next, the fact that those flexible crystal unit cells are just as capable of storing energy as those unit cells in a solid crystal of aluminum sulfate, for example. *(Your attention please dear reader: once again we have two meanings of the word cell. Please see my explanation on a previous page.)* Just because the lipid double layers are fluid and flexible doesn't mean to say the atomic forces in the liquid crystal are different than a solid crystal. Those atomic forces are quite capable of holding information and then passing it on from generation to generation through changes in the DNA of the cell. So it is a beautiful mechanism that was created by God, All That Is, and it works very well and unobtrusively. It is so unobtrusive that your scientists have not fully understood it yet, in fact, most of them hardly understand it at all. Malcolm, you can help with your description of this process.

That is all we wanted to tell you today. We think that is enough to get you started on Chapter 4, because you need to describe the physical set up first, the occurrence of the double lipid layer. Show people this is like a liquid crystal and explain what liquid crystals are, and then explain the memory passed on from generation to generation in our words if you wish.

We thank you Malcolm for doing this once again. We were pleased that you were able to put together the time and the equipment needed to make this recording so that it could be retained and used in your book, in our book we should say, because we are sure you realize that this is a very combined operation of different levels which in one way represent the different levels of memory. There is brain memory and cellular memory. You could say that there is an analogy for our relationship in that we Angels represent the brain memory and you, Malcolm, represent the cellular memory. It's a rough analogy but it could be used to describe different levels of memory and how they operate among the cells of the human body.

So we wish you well, Malcolm, and thank you for this work you are continuing to do with great enthusiasm. We assure you your health is good and will continue to be good so that you can continue to do this work for us. You have no need to worry about any parts of your body being poisoned or damaged. If you live your life with care we will ensure that all works well for you so that you are able to do this work and produce the books that we have ready waiting to be transmitted to you. We will ensure that you have the facilities you need to do this work, like a place to live. You have no need to have any concerns about your physical existence, because we are looking after you and making sure our writer has all the things he needs to do this work and also enjoy his life. Right now, you love doing this work and that is a great driving force because you give intent every day to do this work for us, and we appreciate that and reward you with a secure life situation and good health. The health experiences you have had over the last two years have just been there you make you stronger and to prepare you for the work ahead. None of it was meant to frighten you or take away your abilities or your sensitivities. All those apparent diseases that came to you were there to strengthen you and give you reassurance of our continued support and healthy lifestyle. For your part, all you need to do is feed your body well and to exercise as you have been doing. If you maintain that level of activity then you will be rewarded with good health and clear ability to do the work that we have undertaken with you. We thank you for helping us with this work. We know you love doing it and we assure you we love working with you on it. We will continue our great partnership as long as it takes to produce all the books that we have ready waiting for you to write.

Now we come to the end of our communication for today. We send you our love and very best wishes for this coming exciting year. You will see many things happening that have been predicted and you will enjoy living through this year of great change. You need not have any fears of anything physical happening to you provided you take normal precautions. Don't do anything dangerous to your health like riding motorcycles – with reference to our lady that acquired the heart – a little joke if you like, we know you like to laugh. We love to laugh as well, we all laugh together. Our faith and our strength increase when we laugh, especially when we laugh together. A sense of humour is a wonderful gift from God, All That Is.

Now we thank you Malcolm once again for doing this. We send you our blessings as we bid you goodbye.

Elapsed time 40 minutes.

Malcolm K. Smith / January 14, 2012

Appendix 12:
Angel Communication, February 17, 2012

Greetings Malcolm. Wonderful that you could get your recorder ready so quickly in all this mix up of arrangements in your new house. We appreciate your speed and determination to get it done. We thank you for preparing yesterday to do this. You could tell that this session with us was about to happen and you went and got new tapes and found your recorder. That was very sensitive of you to detect that. We are sure your Guides readily join in this work as this morning when you first got up you wanted to make tea but they said "No, don't make tea just come and talk with the Angels, we have a message that they are ready to talk." So everything worked out well and we are ready to talk to you about the next chapter in our book which you will call chapter four. This is about the cosmic lattice. How those other forms of psychometry are practiced, you already have some notes on that from the previous times, and you will need to refer to those as you write this chapter. We will give you an outline now that will guide you as a framework to put those words from previous times together. The main thing that you need to recall is that there are several methods of reaching the cosmic lattice, like the flower-guided method, the sand impression method and the coloured ribbons method, but they all involve linking with the reader's Golden Angel – Mikael in your case – who makes the connection. You have written about this in your first chapter, on what we will call standard psychometry, and that is the basis for several methods including the three just quoted. They all involve the assistance of the Golden Angel of the reader to make the connection with the cosmic lattice.

Most humans use the cosmic lattice unconsciously – we've talked about this before. In the unconscious use of the cosmic lattice it seems you just wish for things, good or bad, and you get whatever you wish for. That is the basic unconscious use of the cosmic lattice that most humans practice without being aware of it, because it's unconscious. But when a human needs to make contact and does so consciously, then he or she enlists the assistance of his or her Golden Angel. That is the way the connection is made. This is because it is difficult for humans to make the connection consciously, although they can learn to do it. Many people who are proficient at telepathic communications are able to make the actual connection with the cosmic lattice themselves, with or without the assistance of their Golden Angel. You are getting to the point where you can do that, Malcolm, but you haven't practiced it, but that's alright we can use the method that we have established with you, which works quite well. Those other connection methods can be practiced later as we do more of this work.

Let us choose the flower method to talk about in detail first. With this method, the flower is connected into the cosmic lattice which holds the flower's growth patterns, the flower is in communication with all other flowers and with all other living things on the cosmic lattice. As you know, it connects all living things which include rocks and things you probably would not think of as living, but nevertheless they are. All the living things are connected into the cosmic lattice because they helped create it for this planet. Each particular species of plant has a particular part of the cosmic lattice. In your talks, you refer to Rupert Sheldrake's idea that in a sliver of wood cut from a willow tree are all the knowledge and designs that are necessary to build another complete beautiful willow tree. That is right. It seems that the connection to the design information for the tree is in that piece of wood itself because that is what grows. But it's the connection between the piece of wood and the cosmic lattice, especially the part for willow trees that matters, because the information has to flow through it when the cutting is planted.

The flower used for assistance in psychometry has a connection to the cosmic lattice and that's what the Golden Angel uses to find the information required by the human reader. For example, in the psychometry lessons which you, Malcolm, talked about in the introduction to this book, Hazel brought a piece of bamboo which was the plant that you chose to work with – although you did not know that it was Hazel who had brought it. That bamboo plant led you to connect with your Golden Angel Mikael, who entered the cosmic lattice via the plant's connection with the part of the cosmic lattice that deals with the affairs of bamboo plants. When Mikael and the bamboo part of the cosmic lattice were connected, Mikael was able to use that connection to go into another part of the cosmic lattice that deals with Hazel – the "owner" of that particular piece of bamboo – and Hazel's life and Hazel's likes and dislikes. (My Guides tell me that there was an overlap between the bamboo part of the lattice and the part related to Hazel as owner of that particular piece of bamboo. Mikael was able to detect the overlap and so made the connection between Hazel and the bamboo – MKS.) One of the things that Mikael found out about Hazel was that she liked aromatherapy, so Mikael brought that back to you, Malcolm, and you perceived the concept of aromatherapy in connection with Hazel. You announced to the class that the owner of the bamboo liked aromatherapy – among other characteristics – and Hazel revealed that she had brought the bamboo and she really liked aromatherapy.

And that's basically how it works, the plant provides a way or a route into the cosmic lattice. From there, the lattice can be explored fully and connections made with the apparent owner of the plant – we know Hazel had bought the bamboo only a few hours before the reading, nevertheless it stood in for her as a way of connecting to her special part of the cosmic lattice. There, information concerning Hazel was stored, and Mikael could select some suitable piece to bring back to you, Malcolm, so that you could do a psychometric reading. In that case it worked for a person, but of course you can get information on many factual areas of the cosmic lattice by using a plant as an entrée into the lattice.

The cosmic lattice is accessible to all the Golden Angels, they are quite familiar with exploring and getting information from it. They don't really have to explore it because they know just by thinking about it where all the different parts are. As we told you before, if you wanted to know something about strawberry jam you only have to think of strawberry jam and if you are an Angel you go immediately to the part of the cosmic lattice concerned with strawberry jam. If you are a human it may take a while for you to find it because you are not practiced at knowing all these things. If you get the help of your Golden Angel, then he can show you the way.

There are several other psychometry techniques similar to the plant method. There's the sand technique, in which people push their hands into a bed of sand to make an impression. That is another route that the Golden Angel can follow to make the necessary connection. There is another technique using coloured ribbons in which the person who is to have the reading – we will call that person the "readee" – puts the ribbons in an arrangement of colours which is pleasing to them. In doing so, a personality stamp is made; you have heard of energy stamps, well in this technique the readee puts a personal energy stamp on the arrangement of colours of the ribbons, putting on top the preferred colour, putting underneath that the second preferred colour and so on. The arrangement of colours signifies to the Golden Angel of the reader an energy stamp relating to the personality of the readee. In this case, and that of the sand impression, the Golden Angel does not need to go through a natural object to get to the area of the cosmic lattice relating to the person being read, instead the Golden Angel goes directly to the lattice area relating to the readee because that person has established a personal energy stamp which itself acts as a route for the Golden Angel to connect directly with the lattice area concerned with the readee's life. So that is a slightly different technique. All these methods are working for the same purpose and that is for the Golden Angel of the reader to make a connection with the readee's personal part of the cosmic lattice.

To summarize we can say the purpose of all these indirect methods, such as reading information about the owner of a plant or an impression in the sand, is to connect with the readee's special part of the cosmic lattice which is concerned with the readee's life and preferences. One method is to use a plant where the route into the cosmic lattice is that plant's special part of the lattice. Another method is to go directly to the readee's special part of the lattice. In that case, the readee establishes a personal energy stamp by putting an impression of his or her hands in sand or by arranging colours of ribbons.

So that is quite an efficient method that has worked for thousands and thousands of years on the planet Earth. People who were viewed as seers used these methods in a natural way and weren't interested to know how these things worked, because in those days people did not think about things scientifically. Even when they did start to think scientifically it was about physical objects – for example why cannon balls of different sizes all fall at the same rate. It wasn't until Descartes wrote "I think therefore I am" that any questions were asked about humanity's methods of connecting with each other or with the universe at large. That's when people started to become curious to know how these things work, especially people like you, Malcolm. You were born curious and that was set up deliberately because, as you were told by your Guides, you were trained by us before you came into this lifetime to be a very curious, questioning person because you felt you had to understand God's plan for the universe. That is a very worthy attitude to have and it brings you to the point where you can find out many things, explain them to your fellow humans through our books and do a great service. We realize that there are not many people who are curious about these things because they come into the Earth plane with curiosity and a wish to understand the Earth, but they lose their train of thought and very quickly become totally immersed in physical things. That is a pity, but things are changing in that respect.

You notice in your school workshops some young people are really exploring and are quite fascinated by the universe they find themselves in. In the last week or two you have had some very penetrating questions from young people that show they really think about their existence. We bring to mind the question asked by a young boy this past week, after you described what an atom was, he asked, "How many atoms are there in the universe?" We know you couldn't answer that question but you made a valiant effort to explain it in terms of the Avogadro number which gave him an answer he could take away and think about and maybe lead to some greater understanding of structure of the universe – this wonderful thing that God has created for us.

We Angels of Crystal Light and you Malcolm together are providing understanding and information to those who will listen, who want to hear about these things. This is a worthy task that we enjoy taking part in and we know you love being part of this work. We thank you for that Malcolm because you are our communication person, our contact person or the "go to" person – those are the terms used in your modern world – on the Earth for all the understanding of the lessons in how things work in the Universe. This is a great task that we are very happy to be part of and we are sure you are extremely happy to be part of this as well. We will work together to produce our second book and many more and you will become known for this kind of work. People will call on you to talk and explain a lot of things that are not obvious by normal scientific methods and beliefs that you have on Earth at the present time.

Now there are the first instances of scientists realizing that it's important to take into account God, His creation of the universe and the love that flows throughout that wonderful universe. Without understanding of that love, scientists cannot pursue their investigations that are concerned with only physical things. It is not possible to explain the universe unless you take into account God and the love He has given the universe. That love is the essential driving force of the whole universe as humans perceive it. Without love it is not explainable. But you have made a good start in our first book in putting that value of love forward as an important scientific "variable" if you like. You said in the introduction to our first book that "Science doesn't make sense without God." We hope that as a result of the books that we are going to write that lesson will come through and scientists, and all people on Earth, will understand that the universe and its structure, its meaning and its purpose are only explainable in terms of God, All That Is and the wonderful love He showers upon us all.

That's all we have for you this morning, Malcolm. Not a very long session and we really appreciate the fact that you responded quickly to the call for communication between us. We thank you for doing this work, for being so responsive to our call. We love you and we bring you all the blessings that are ours to give. Thank you Malcolm. Go and enjoy another day in your new apartment, in your new world where you have a new view on what is happening in the world, a new way of interpreting things that you see around you and in thoughts that come to you. We are very happy that we can work together in this way to help all the humans yet unknown on planet Earth that are about to incarnate or have just incarnated or are waiting in that bardo between physical states. We are here to help all spirits that become human or who have been human at some time. The knowledge, the understanding that we are creating between us is a very large part of that education process.
We now leave you with a great thanks, with all the blessings that are ours to give and a great love that spreads over the whole Earth and all humans. Goodbye.

Elapsed time 35 minutes.

Malcolm K. Smith / February 28, 2012

Appendix 13:
Angel Communication, March 8, 2012

Hello Malcolm. Great to be with you again. Thank you for agreeing to do this once more so soon after you had finished writing Chapter 4. We are making good progress and we are very pleased with the way things are going. We wanted to come to you now to take advantage of the good momentum that you have built up and get you prepared for the next part of the book, which will be Chapter 5. As you guessed, this chapter is about homeopathy and impression of energy patterns onto molecules in medications and things like that. So that is the area that we want to talk to you about today, as you guessed.

We want to talk to you today about homeopathy and psychometry. That could be a good chapter heading for this section which should make people very curious. We want to start by talking about the structure of water. As you know from Dr. Emoto's experiments, water can be influenced by energy patterns and human thought. Normally water molecules exist in clusters called oligomers which are bonded together by hydrogen bonds. Water oligomers occur in a range of sizes consisting of up to 1,000 or more water molecules depending on the natural conditions of the water. Spring water is more beneficial for humans than stagnant water because in rapidly moving natural water the oligomers are relatively small (about six molecules). If the water is stagnant, and especially if it becomes polluted with negative energy from either human or natural sources, then the water molecules are slower moving because the energy is low and they tend to collect together into relatively large oligomers consisting of thousands of molecules. However, under the influence of human intent which is expressed as good or happy thoughts, good wishes and prayer – any of those forms of human thought – the oligomers are constrained into smaller groups of the order of six molecules. As you know from your structural chemistry, six molecules bonded together as hexamers are the most stable configuration, that is why snowflakes have six-fold symmetry.

The stable hexamer structure of water provides the most efficient use of water required by plant and animal life on this planet. Water is the liquid of life on Earth, your human bodies consist of at least 70 percent of water, depending on which part of the body you are talking about. The average is about 75 percent. The water in human, animal and plant "bodies" has to flow through very small orifices in cell walls, and if the water is in very large oligomers its apparent viscosity is too great for efficient flow in and out of the cells. If you look up the dimensions of orifices in the cell walls you will see that they are slightly larger than a water hexamer, so that hexamers can slip through the orifice easily, the apparent viscosity is low and water flows through the cell efficiently. On the other hand, if the water is present as oligomers consisting of about 10 or more molecules then they tend to get stuck in the orifices. If that happens, the apparent viscosity of the water is high and the flow of water through the cell is slow and inefficient.

Water in the form of six-molecule clusters called hexamers fit through the cell wall orifices easily and good flow conditions are experienced by the human, animal or plant. Under these conditions, the water is experienced as beneficial to the life of the cells and the whole organism thrives on the hexamer-based water, because water flows through the cells quite rapidly and everything works efficiently. When stagnant water is used, and that includes water that has been exposed to negative influences, then the water is in relatively large oligomers, and this situation leads to stagnation of the water in the cells, because the large oligomers become tangled and take a longer time to move through the orifices because the apparent viscosity of the water is increased. Multicellular organisms like humans, animals and plants don't thrive on water containing relatively large oligomers. (As an aside we can say that if you send love to your plants they will thrive because the love that you send equals good intent which is part of the love energy that we talked about in our first book. [Ref. 5-2] Love energy puts the water molecules which are already in the bodies of the plants into the most stable and beneficial form which is hexamers – or smaller clusters – throughout the plant's structure. The water flows through the plants very readily without any blockages and the plants are "happy" and grow strong and beautiful.)

Most of Dr. Emoto's work was based on the perfection of ice crystals that were formed from water from different sources. Perfect ice crystals were formed from "good" water – that is either fresh spring water or water that has been blessed by humans so that the water consists mostly of hexamers (six molecule clusters). The beautiful ice crystals from good water mostly had the same structure as natural snow flakes with their six-fold symmetry. Ice formed from polluted or stagnant water appeared to have a non-crystalline structure that resembled congealed mud. You, yourself, Malcolm, did some experiments in which you played beautiful baroque music to water for 12 hours and you also played some discordant rock music for the same time to another sample of water. Then you did some wicking experiments in which you measured the speed at which water climbs up filter paper strips as indicated by the movement of a spot of coloured dye solution placed on the dry paper. You showed that there was a consistent physical difference between water treated with hard rock and gentle classical music.

What we wanted to tell you, Malcolm, was about homeopathy preparations, and this is in addition to the discussion of oligomers, hexamers and the purity of water influencing the life of plants. In addition to that, energy stamps can be put on water, and these energy stamps come in a variety of different frequencies of love energy. Basically, they are a message, the energy stamp contains information that is different for different applications. Using an analogy that we referred to before in a previous talk about this, the energy superimposed on the molecules puts them in a pattern. These patterns may consist of hundreds of thousands of molecules – it's a much larger scale effect than the hexamers and the things we have just been discussing. The best analogy we can describe here is that of crop circles – the dimensions and the detailed structure of the crop circles are an expression of information in a message. Very often, those messages are meaningful to humans who see the crop circles. The crop circles are formed by an energy stamp being placed on millions and millions of wheat stalks so that they become visible to humans, and humans understand at some deep emotional level the message of the crop circles – which are not circles anymore, they are crop formations, that's the word we should use. But it is a very big scale event that occurs, much larger than six wheat stalks being put into this energy stamp. So the Emoto effect and everything we have been talking about up to now is on a relatively small scale of up to 1,000 molecules.

Whereas when we come back to homeopathy, the energy stamp that is put into water to form the equivalent of a crop formation in the water molecules involves hundreds of thousands, and maybe even millions, of water molecules. But there are a lot of molecules in a glass of water so there's no problem with the numbers here. You only have to think about the Avogadro number analogy that we talked about last time to see there are plenty of water molecules to be formed into – we could call them crop formations – but these are molecular formations.

We can draw an analogy here between crop formations of millions of wheat stalks and the situation that occurs in a homeopathic preparation where an energy stamp has been put on the water molecules so that the water molecules form up to very large energy stamps. They are not constrained by hydrogen bonds, so it is not an enormous structure that has difficulty in moving through the water. It's just that the intent that is imposed by the energy stamp on the water molecules makes them want to be in a form that the energy has expressed by the structure. Individual molecules in the structure are quite free to move about and pass through cells in bodies of animals and plants, but in moving through the cells, they release that energy that is imposed on them. It is difficult to visualize the structure that is imposed by the energy stamp on the water molecules. It's a memory effect. The molecules of water in homeopathic preparations have a tendency to form into structures equivalent to crop formations, but it's just a memory effect. An energy stamp is quite fluid, it leaves a tendency to form a particular structure but it doesn't have to stay in that structure as it works through a human body.

We give the analogy of a tape recording of a piece of music. The energy of the music is imposed on a length of tape, a complete symphony can take up a complete tape, and the energy is dispersed along the magnetic iron oxide particles that are recording the magnetic signal, so that there is some order established according to the magnetic signal, the order is established centimetre by centimetre of the tape. The order is established in the magnetic memory of the individual iron oxide particles in the coating on the tape's surface. Viewed overall, if we could see the patterns that are imposed on the iron oxide particles, we would see a beautiful design that is equivalent to the molecular "crop formation" in the homeopathic remedy. We have an analogy here, the crop formation is a static structure but when the crop is harvested that energy pattern is still present in the crop, in the wheat ears we will say, and that energy pattern goes into the preparation of bread, beer and food in general. Humans absorb the energy pattern and it is beneficial to them; that energy comes into their food and bodies as love energy from which they benefit.

Coming back to the tape recording analogy we have the energy of a beautiful symphony that is imposed on the iron oxide particles of a tape recording. That tape recording appears almost random when it's viewed on a small scale, but if you could view the patterns of energy in the whole recording from one end of the tape to the other, you would see a beautiful pattern which represents the music that is recorded on the tape. In the same way, the energy stamp that is recorded on the molecules in the remedy is spread out over millions of molecules and if you could view the pattern from a distance, you would see the beauty of the homeopathic remedy – the beauty of the fluctuations, the patterns of love energy that are recorded in the molecules. The analogy here is that the water molecules record a beautiful, beneficial pattern of love energy which is directly analogous to the magnetic patterns imposed on iron oxide particles on the surface of a recording tape. Another way of putting this is to say that the water molecules in the homeopathic remedy carry with them the message of the remedy. For the remedy to have effect, it is necessary for the whole recording to be absorbed by the patient that is taking it. This is equivalent to saying the whole of the crop formation that is imposed on the molecules of water needs to be absorbed by the person to get the benefit of the energy pattern imposed by the energy stamp.

That is a fairly difficult concept to explain in simple terms, we are having some difficulty in getting the right words from your memory banks, Malcolm. We think we have expressed it enough. As you transcribe these words onto your computer you will find these energy patterns are visualized by your body and it will make sense. You have to look at the whole overall pattern that we are translating rather than concentrating on a few words. The main point that we want to make is that the water molecules are used as a working medium to record energy patterns. Those energy patterns are beneficial to the people taking them as a medication, a homeopathic remedy. It's a different scale of numbers of water molecules that are involved. It's similar in some respects to what we will call the Emoto effects, but those work on relatively small numbers of molecules, whereas the homeopathic remedies work on millions of molecules. The analogy we gave there was the millions of wheat stalks involved in an energy stamp that you call a crop formation, just as the energy stamped into a crop formation can flow into the bodies of humans in the form of food – for example bread and beer – so the energy stamps on the water molecules benefit the bodies of the patients who drink those energy stamps.

Yes, it's starting to come together. We can see several analogies running in parallel strands and we think you are getting the sense of it and you will be able to put this material together to explain it to your fellow humans in the book. Of course if you have difficulty you can come back to us through your Guides and we can explain certain aspects of this if necessary. You will have to clean up the language when you transcribe this recording onto your computer memory, we can help do that as well. We know you are good at rearranging the words, at making them into grammatical sentences.

On this occasion, Malcolm, you were anticipating eagerly the information and you didn't stand back far enough because you were curious to know about the science behind this. We had some difficulty expressing through your memory banks the concepts, but we think we got there eventually, and we can see the patterns developing. When you are recording these communications you should try to let the words come through without understanding them. The time for your understanding is when you are actually transposing them into written form. Just let the words come from us Angels and you will get the sense of it later. We can't fault you for that because we know you are very curious. You want to have this knowledge and pass it on to your fellow humans.

There was one little bit that we could tell you were thinking about, and that was not all homeopathic remedies are liquids – made of solutions in water. That's quite right, some are based on solid material which is compressed into a pill. You yourself took some based on sodium chloride as the recording material. The crystalline structure of sodium chloride or similar materials can be used to store energy – Big Surprise! This is what you dealt with at the beginning of our book in which we talk about energy that is stored in the crystal lattice structure. Coming back to the sodium chloride-based homeopathic remedies, in that case the energy stamp that is put upon the sodium chloride is just like the one that is put on water molecules – it's just a different recording medium. There are several other chemical compounds that can act as recording media, there are some oils and other materials like that, but the principle is the same. An energy stamp is superimposed on molecules, which carry the message into the body of the patient who is taking the remedy. The energy stamp is transferred to the cells of the body that are doing the healing work. We don't think we can get into the actual mechanism of the energy stamp's beneficial effect on the structure of the human body – just let's say that it's like healing done by humans on each other. The healing given by humans to each other is based on spiral love energy known as Chi or Prana and described in our first book. People that practice Qi gong, like Wayne Wia Ling, send energy patterns into a patient's body to heal them. In the case of homeopathic remedies, a similar healing energy pattern has been put into the carrier molecules that are absorbed by the patient. It's just like a Qi gong person sending energy into a patient's body, except it's been put into a molecular structure which is put into a bottle. That bottle is sold to the patient who takes the medication and receives the energy stamp that is healing to that person's body.

We think now we have covered all the aspects, we will let you sort out this explanation as you transcribe this tape. If you have difficulty with any of the concepts you can come back to us for clarification. But we think we have explained each of the concepts and you will be able to organize them into a logical presentation in your chapter.

There is one other aspect that we didn't cover, and that refers to an account in the book by McTaggert (*The Field*) which we referred to in our earlier communication. In the preparation of these energy stamps on water or some other molecules, if a person with very strong mental characteristics – that could be interpreted as negative or positive – is present during the imposition of the energy stamp into the molecules, then that person can interfere with the process. There was an account, in the book by McTaggert we have just referred to, of a woman who apparently spoiled the results of a research project on homeopathic preparations because she had a strong mental influence on them. It was realized that her influence negated the results of the tests that were being done in just that one case. When she was removed from the situation where she could influence the results and all the other people involved had no influence on the absorption of the energy stamp by the molecules, then the preparations were successful homeopathic remedies. That is a point you need to add into the chapter, Malcolm.

So we think we have it all. Not the best transmission that we've accomplished so far but a workable one that we think you will be able to organize when you transcribe it. We ask that you try to put your mind to one side and not be so interested in the information, because in doing so you are interrupting the flow of information. Thank you for doing this once more and for your interest in our work. We know you anticipate the information coming so quickly that you have been able to write another chapter and you are very interested to know about these things personally. But in this case, your curiosity got in the way a little bit, but no harm was done. We thank you for your interest and continued devotion to this cause of us bringing information through you to your fellow humans. Thank you for having tapes ready so when we came with a very quick early call you were able to respond.

We say thank you to you once again, Malcolm. Thank you for doing this work and thank you for all the love that you express through doing this work. We leave you now with all the blessings that are ours to give. We hope you have another happy day and a good run – we know you are dressed ready for exercise. Go and let these words we have given you percolate through your mental processes, and you will come up with a very clear explanation based on analogies between water molecules, crop formations, magnetic recordings and homeopathy. Thank you for all your attention and devotion to this work. We love you Malcolm and we wish you a very happy day. Goodbye.

Malcolm K. Smith / March 21, 2012

Appendix 14: Angel Communication, April 6, 2012

Hello Malcolm, glad you could make it again this time. We didn't give you much notice, but we know you had a good stock of tapes ready, and you were able to put this together at the last minute when you were given a warning last night. So, we are all ready to go once again, and we thank you for doing this once more and for all the good work you are doing in writing our second book, which is of course what we came to talk to you about today. The last chapter is our concern at the moment, we think there should be another chapter after Chapter 6, the chapter on water. Chapter 7 will be the final chapter in the book, and that talks about the processes of making homeopathic remedies and brings together the work on the minerals. The minerals that act as homeopathic remedies in stone – most homeopathic remedies that we have talked to you about are in water. The homeopathic remedies that come as salt and other solid materials – we've said that the homeopathic remedies signal the pattern of spiral love energy – we've developed a name and were are searching in your memory for that name – the energy print(?) can be placed on water, can be placed on solid chemicals such as salt and can be placed on minerals as well. (Long pause) We're still not clear on that name. (MKS: I'm still here – I'll look it up in the book – another long pause).

(MKS: Alright I think we are ready to start properly now. I'm sorry about that delay.) Hello Malcolm, don't worry about these minor delays, it's good to get the word we were looking for – it's "energy stamp." We'll go back and talk about energy stamp feeding into homeopathic remedy preparation. Chapter 6 has been about homeopathic remedies, how they are prepared focusing on water as the carrier medium. Chapter 7 will be on other aspects of homeopathic remedy preparation focused on solid material such as salt. You yourself Malcolm have taken homeopathic remedies based on sodium chloride (salt) as the carrier medium in place of the water, in what we will call the "conventional" type of homeopathic remedy, although conventional is hardly the word, because they are not accepted by so many doctors. But you know what we mean by that – regular homeopathic remedies were mostly based on water to start with. The water has received the energy stamp, putting the molecules into a big cluster or oligomer consisting of hundreds of molecules. We talked to you, and you recorded all that information, about how they pass through the human body and give up their energy, although they don't need to stay in the same configuration as they were in the container of the homeopathic remedy. Nevertheless they can give up the energy as they go through the body, which benefits from the energy which is contained in the energy stamp on the molecules. So we have completed Chapter 6, we know you have not completed the writing yet, but you have all the necessary material down that we gave and you converted to the written form, from the channeled material that we transmitted through you, so you have enough to complete Chapter 6.

What we are starting today is the new material for Chapter 7. We don't want you to modify what you have already for Chapter 6, that is fine as you transcribed it and everything is fitting together nicely. We will just say about it that you need to write a brief paragraph or two about water as a recording medium, and you have some notes already prepared on that so it can stand as it is, then the material we gave you can finish off the rest of Chapter 6. Now we start some new material for Chapter 7, and this concerns the use of solid materials, such as salt, that we were just discussing, and also the use of minerals like the black tourmaline that you sent to Brenda when she was first diagnosed with cancer. You made her a necklace of black tourmaline because you found in your healing mineral book (Ref. 7-2) that it helped support the body's own healing power during cancer treatment. She wore it during her operation which was very successful and the black tourmaline acted as a homeopathic remedy for Brenda and helped cure her cancer. This is the sort of situation that we want to talk about now.

You already have some material in a previous communication from us about how minerals – and all chemicals – give off energy that is characteristic of the structure of the chemical. You know about these things, Malcolm, with being a chemist, that's one of the reasons why we work through you because you understand terms that we use. Any chemical has a characteristic energy that is measurable. For example, you used to measure, over a range of light wavelengths, the optical density of solutions of chemicals – we would refer to that in the common world as the colour. The colour spectrum that was received by the human eye is the signal that the particular chemical gives off. All materials give off electromagnetic radiation which is due to the movement of electrons in the molecules, but not all materials give off radiation in the visible range, and humans see those that do not as colourless. Nevertheless they are giving off a signal, and if you have an instrument of human devising such as a spectrophotometer that can measure energy given off at all wavelengths of interest, then you can pick up a signal.

An example we would give is: in the ultraviolet range of the electromagnetic spectrum, humans are not able to see anything visually, but you are able to measure it and so detect a signal from the chemical, a characteristic fingerprint, we will call it, of the chemical. That fingerprint is the "song" of the electrons – or the characteristic tune the electrons play with electromagnetic vibration – that arises in the movement of the electrons that bonds the atoms together in a particular kind of molecule. That "song" is listened to by your instrument, which can detect a particular chemical by comparing it with patterns of vibration made by known chemicals – songs that have been sung by known chemicals – with all the structural analysis that goes on behind the scenes in this science. You can then put together an absorption, or transmission, spectrum of electromagnetic signals given off by the molecules, and from the spectrum you can identify the chemical.

When you were doing your Ph.D. research, you would measure the light radiation given off at many different wavelengths by one of the compounds you were studying, and you would say, "Ah yes I recognize the shape of that curve, it's the curve of anthraquinone" (MKS: a chemical involved in making dyes), if that were the case. That is associated with your development with the help of your guides of a device that you talk about in our first book (Ref.7-1): A "micro-vacuum cleaner" for collecting very small samples such as whisker crystals that grew from the surface of the dye films you were studying. You were able to collect some whiskers and dissolve them in alcohol to make a solution, and then when you measured its absorption spectrum, you had proof that the crystals were made of anthraquinone only and no substituted (modified) anthraquinones were present in the whisker structure. That is an example of the way the electromagnetic song of the molecules can be of use in identifying them.

Scientists are able to use the electromagnetic message that comes from atoms and molecules to identify materials in other parts of the universe. For example, in the case of the planet Mars at the present time and over the last few decades, there has been a lot of research looking for water. The research uses the characteristic electromagnetic song of water molecules to tell the scientists whether there is any on the surface of the planet. This was during the time that humans could not send machines to Mars – we know that they have now – but before that was possible, this was the only way they could determine whether there was any evidence of water on Mars.

This method of molecule exploration has been extended to other planets and star systems. The colour of the light – measured over a range of light wavelengths – that comes from other planets is the same as the thing we are calling the electromagnetic song. This electromagnetic song is broadcast throughout the universe by the molecules on another planet – let us say Neptune, for example – and scientists on Earth can receive that light signal, that electromagnetic song, and by analyzing it determine the composition of molecules on the surface of Neptune. This is an extremely useful tool in research and is one of the ways humans are starting to understand the structure of the universe in which they find themselves. It is a great benefit for humans to be able to measure these electromagnetic song signals and it has helped make understandable the place of humans in the universe and how the universe around you relates to your home planet Earth. In time humans will go to other planets, they are already starting their exploration of the Moon, Mars and Venus. Very soon in human terms, scientists will be sending vehicles to other planets and eventually they will travel themselves to those planets, just as they travelled to the Moon – Neil Armstrong's "giant leap for mankind."

Scientists of the 18th and 19th centuries realized that all molecules give off a characteristic electromagnetic song. By tuning into that song, they can obtain information on the composition, structure and what atoms are involved in materials that are lying on planets millions of miles away from them, that is a great benefit in understanding the origins of humans and the place they occupy in the universe.

But of course those electromagnetic songs are not only of use in space exploration and analysis of chemicals on surfaces of other planets and stars. They also have great value here on the Earth because the chemicals are still giving off their characteristic song even when they are lying next to a human's skin. That's how the minerals work – like the black tourmaline that helped to heal Brenda's cancer. They give off their characteristic electromagnetic song, broadcast it to any body – we mean the body of any person or animal – that is near and that can be beneficial, very beneficial for a particular disease. We know that when a human body has a particular condition, maybe for example the one you call cancer, then certain electromagnetic songs are able to heal that medical condition – we are talking here about cancer, but of course this applies to any medical condition of human or animal bodies.

Over many centuries of experience, humans have come to realize that certain minerals have beneficial properties for their bodies, especially if the mineral is close to the body or even in some cases, taken into the body in the form of powders. This is because the electromagnetic song is like the energy stamp that is put on water in homeopathy. In fact, in some cases that is the way the electromagnetic song is delivered to the body, by mixing a certain molecule with the water, so that the energy stamp from the material, which is put in as the "seed" chemical, is recorded by the water's structure and is transmitted to the body of the person taking the remedy. That is what we talked about in Chapter 6, now in Chapter 7, we are going to talk about how the material that is the originator of the energy stamp is able to be directly influencing the human body's health. The energy stamp is ever present in the molecule that is to treat the human health condition, if the treatment is not to be through water, then through some other solid material such as salt, or it may be conveyed directly from the originating molecule right into the human body concerned. So here we have a gradation of treatment methods that we will enumerate.

First of all, we define our terms. The healing (or seed) molecule is the term we use for the material that brings the appropriate electromagnetic song to the molecules in the human body that require help to recover from the condition known as disease. The original healing molecule sends its electromagnetic song into the receiving molecules of the human body, and that song changes the structure of the receiving molecules and the cells that contain the receiving molecules; in effect a healing takes place. The healing effect is complex and we can't define it any more (MKS: However it is discussed a little more at the end of this appendix 14), but we don't think it is necessary to because it is understood by humans that when healing has taken place, the structure changes in a way that you talked about in the first book (Ref. 7-1). So this conveying of the electromagnetic song is the all-important step and it can be done in a number of different ways:

- It can be done by imprinting the energy stamp of the healing molecule onto water and that's what we talked about in Chapter 6.

- It can be transmitted through some other chemical substance, which is usually a solid because there are not many liquids that are suitable for use as the transmitting material other than water, of course, which is a prime example with its strong ability to record energy stamps. Mainly in this second category, the transfer of electromagnetic song is through a solid carrier material, such as sodium chloride, in which the song of the healing molecule is recorded. The soluble sodium chloride is taken by mouth into the body of the person to be healed, and the energy stamp is delivered to the parts of the body where the treatment is required.

- The third method is direct application of the healing molecule to the human body. On rare occasions, this is done by grinding up the healing material and putting it in tablet form, which is then put in the human body where it goes to the required location and delivers its electromagnetic song of healing. But more often, the electromagnetic song can be broadcast from the surface of the body into the area where the treatment is required. This is the case of the healing molecule being in a material that you would call a mineral or maybe a semi-precious stone and, through many years of practical experimentation, the electromagnetic song of any particular mineral has become recognized as being efficient in healing a particular disease. That knowledge is contained in many books such as the one you have, Malcolm (Ref. 7-2), on the use of minerals to heal diseased conditions of human or animal bodies. You have used that book to find the appropriate stone, which was black tourmaline, that you made into a necklace for Brenda to wear and help heal her cancer. In those cases, the stone is put in close proximity to the body that requires the treatment and the electromagnetic song goes directly from the healing molecules straight into the body – we would say the song is broadcast into the body – and the body responds by receiving the signal and being healed.

It is a convenient common human practice to take those healing minerals, confer on them titles like "Healer of Cancer" and fashion them into jewelry that can be worn on the body. This is a convenient way of broadcasting the message from the healing molecule into the body – and it works very well, as you know from Brenda's experience. It was not accidental that Brenda's need came up at that particular time, it was planned that there would be a demonstration of the efficiency of jewelry of appropriate mineral content in healing. This was set up for you before you knew about this particular book, but now looking back you can see the point of having this demonstration. So that is how minerals and semi-precious stones are able to heal human diseases. It's like homeopathy, except there's no intermediary material to record the energy stamp. In this case the energy stamp is delivered directly from the healing material directly into the body by a broadcast of the electromagnetic song. You could say it's like having the singer of the song broadcasting the healing song directly into the human body rather than a recording being made in water, salt or some other material and then playing that recording inside the human body. So there's an explanation for the thousand-year-old folklore which tells which minerals are able to heal certain human diseases. That's all set out in encyclopedia of healing stones like the one you possess, Malcolm (Ref. 7-2).

We think that's rather a good analogy: molecules doing the healing are singing their electromagnetic song all the time. We can record that song in water or salt, deliver the recording into the body or apply it to the body in some way. Alternatively, the healing molecule can be close to the body in the form of jewelry, and singing its song all the time, so that the person who wants to be healed can wear the jewelry that emits the healing song, and receive the benefit of the electromagnetic song of the healing molecules directly into their bodies. It's a very efficient method that works well.

Regular doctors tend to laugh about this sort of thing, so do conventional scientists, because they can't see how the electromagnetic song can possibly heal a human body because they are so fixated on the idea of having to put chemicals into it. Some people in responsible positions are beginning to understand there are more ways to deliver healing energy stamps than has been realized in the past. In what has been regarded as superstitious knowledge and associated with witchcraft and psychic charlatans, there is in fact a basis of science. It is a new science that is not generally recognized but some people responsible for health issues like this in the human situation are beginning to realize that there is more to the folklore on minerals for healing than has been acknowledged in the past. This is marking a new trend – you can refer to some of these people. In the book, *Tuning the Diamonds* (Ref.7-3), there is mentioned quite a few people who are working in this area. Also in the book called *The Healing Power of Water* (Ref.6-3), put together by Dr. Emoto, there are a number of people who are active in this area. It would be good to give references to those people and their work at the end of Chapter 6 in the new book.

That is the final message as far as we know, unless you have any questions when you come to transcribe this or when you come to write Chapter 7. This is also the final channeling that we expect to give on this particular book. But don't think that this is the final book by any means, there's a lot more work that is coming for you. We will be starting up channeling for the other books quite soon. But for now, we have completed our intended channelings for our second book. We want to thank you, Malcolm, for undertaking this work and making it possible for information to be broadcast in this way, information which is going to be beneficial to many humans as they see the possible mechanisms through which these strange old methods appear to work. There's a great body of knowledge about the beneficial effects of stones and minerals worn as jewelry or touch stones held close to the body, but no one has come forward with an explanation for how that beneficial healing process works.

Perhaps we should say a little more about the actual process of the healing. The energy that's given off by the healing molecule – the electromagnetic song we called it – is received by the disease-centred molecules and received by the cells that are around the disease and maybe carrying the disease. The healing song is received in the form of electromagnetic radiation which is at such a high frequency that you would call it spiral love energy. It is there in the healing molecule in addition to the regular frequencies of the electromagnetic song. The electromagnetic song consists of layers of frequencies. The frequencies that you humans measure as the electromagnetic song – the emission or absorption spectra – the electromagnetic radiation in your normal range, is accompanied by overtones, the frequencies of which occur in a fractal pattern. They are fractal frequency overtones that range from normal frequencies of electromagnetic radiation that originated in the movement of the molecules right up to the frequency of spiral love energy which you know is 10^{33} Hertz. When an electromagnetic song is broadcast via a mineral into a human body, the electromagnetic part of the broadcast is accompanied by fractal overtones which do the actual healing. It's the same sort of energy that occurs in the energy field you develop when you combine two vibrations, like in your walking meditations you combine *yin* earth energy and *yang* sky energy to create a scalar energy field, which has frequencies that include spiral love energy. Those same frequencies corresponding to spiral love energy are present in the signals of the healing molecules and in the minerals that are used as healing jewelry. We think you will understand when you transcribe this how that energy works to heal. That same energy of very high frequency is present in the energy stamp that is conveyed by the water or solid homeopathic recording material into the body. So, the electromagnetic song that is picked up by your instruments is just an indicator to the people on Earth of the presence of particular molecules on other planets and stars, and is used as a method of analysis. But accompanying those frequencies that humans regularly measure are other frequencies corresponding to scalar energy fields and spiral love energy at frequencies of the order of 10^{33} Hertz. It is those overtones which occur in the whole spectrum of frequencies emitted by the healing molecules which relate to each other in a fractal way. It is the higher frequency component of the spectrum that is the actual healing energy. It is the same kind of energy that is summoned by a Qi gong master – like Wayne Wia Ling – and directed into the body for healing purposes.

We think we have explained that sufficiently that you understand the concepts now, Malcolm. If you can't put it into words for the moment, you have words from us which you will understand when you transcribe them – you will understand the basic mechanism, operation and release of homeopathic energy stamps and the electromagnetic song of molecules.

So now we leave you Malcolm, we think you have everything you need for Chapter 7. Maybe you should call it something connected with electromagnetic song – it is a simple device that will resonate with people, but don't make it too scientific so they can't understand it. Thank you once again for doing this work with us, we know you love doing it, we hear you say so in your prayers. We are very happy that you continue loving this work, we love you just as you love the work. We give you our thanks and appreciation for all the good work that you do. Now it's time for us to leave.

Don't discard the search at the beginning of the tape for the term "energy stamp" – that's a good illustration that there are problems that can occur in this channeling process. You overcame this problem briefly by looking up in your past records, which is a good reason for having readily at hand a complete record of all the channeling work. It brings instances like that into play and makes it possible to find the answer when it's needed. We were having difficulty in finding that term in your memory banks and the written word came to our assistance and to your assistance, and made it possible for the whole thing to work today. We were very happy that we were successful, you were able to stand aside during this channeling much better than during the previous one and we were able to deliver the material that we had prepared. We thank you and give you our blessings and leave you to enjoy another lovely day on the Earth plane, Malcolm. Thank you, we love you. Goodbye.

Malcolm K, Smith / April 21, 2012

Appendix 15:
Angel Communication, May 28, 2012

Hello Malcolm. My, what a long time it took to get you ready for this morning. We understand that your medication makes you very sleepy and that had to wear off. We've just got you in that "inbetween" state where your senses are still dulled a bit by the sleeping pill, but you're alert enough to receive our messages, so that will help the flow. Just relax and let the information flow as we pass it to you.

Today, we want to talk about ghosts and spirits because that was the question you had in your mind. You feel that there is something connecting you with this ghost state and psychometry – you are quite right, there is an allusion between the two. You could say that they are the positive and negative sides of the same influence. In the psychometry situation, as you know from what we have been describing, the information energy is stored in the interior of crystals. In crystals, we can imagine little buckets of space formed between the atoms – what scientists call unit cells – and the information energy can be stored in the little buckets of space in the crystal. We have gone through all of this earlier and made an analogy with musical notes stored on a three-dimensional staff, where the information is stored in each measure. The notes represent little bundles of information energy that are being stored, because contained within them is information about pitch, time and all the other instructions for recreating the composer's concepts. Just as you can read out music from the measures, so you can read out the energy from the unit cells in crystals.

When you do psychometry, your Golden Angel (or you) goes into the crystal structure and reads out the energy, finds information that is stored there and brings it back to you. You form it into concepts and ideas about the owner of the piece that is being psychometrically read.

Well, a similar thing happens in a natural way in materials that have crystals embedded in their structure. As you know from your work, crystals are most commonly found in metals and minerals. The houses on Earth are mostly built of materials: stone blocks, bricks, cement and wood, all of which have partly crystalline structures and so there is a plethora of materials suitable for storing information energy. Over long times, the events in the structures that are made of these materials get recorded in the crystals through a natural process. It's always a natural process – there is no storage of information by humans in any of the situations we have talked about in this book. It always happens through a natural mechanism where the building and the materials – and we should include textile furnishings in houses as well – all these materials absorb the information energy into the structure. If it's a very old house or a castle, then there's a background of energy existing in that place, which may be very peaceful and which has been absorbed into the crystalline materials of the building. When you walk into such a building, you feel there is peaceful energy in this place, but what is really happening is you are reading the peaceful information energy that's been accumulated over many years in the walls and the structure and textile furnishings in the building.

But on occasions, there's some energy that is dissipated in the building which is far from peaceful. Situations such as the sites of big battles fought in wars or places of torture in the old days, places where murders and other dreadful deeds have been committed or other personal harm has come to some people, the energy from these events makes its way into the crystalline storage of the materials of the building's structure. Energy pulses from these events go into materials that have crystal structures suitable for storing them and the result is a terrific batch of energy stored in the fabric of the building. When sensitive people come into buildings where bad deeds have been done, then they are often aware of the energy that is stored in the fabric of the building. They read that energy out and understand it as an atmosphere of the place.

You yourself have been to Culloden in Scotland where a great battle and massacre happened. In that place, there are a lot of crystalline minerals in the ground and the energies of the terrible deeds that were done that day were absorbed into the underground minerals. Sensitive people that go to the battle site in recent years immediately become aware of the terrible atmosphere that recalls what happened there in 1746, when so many people were killed. The animals and the birds in that area are constantly aware of the bad energy in that place and they stay away from it. That is why when you went to Culloden, you didn't see or hear any birds in the air or in the trees and bushes. Actually, it is a place to be avoided because of the bad energy and the birds are very sensitive to that and they stay away completely from the area of Culloden.

So, that deals with atmospheres that we sense in some places and that is a relatively low level of recorded energy. But, in some situations where there is a very terrible stressful event, maybe in the way someone is killed with a lot of mental anguish involved or responsibilities that are not completed at death, those types of events can lead to storage of a much stronger energy, in fact we would say it's not so much stronger energy as the sheer quantity of energy which stored in the material and fabric of the place or buildings. When the bad energy reaches such high levels as it does in these places, then it doesn't wait for a human to read it out, instead it appears spontaneously, it expresses itself as a strange event that you refer to as a ghost. The energy of the ghost only comes out of the crystalline part of the fabric when human attention is directed towards it. If a human comes into a room, and in the fabric of the room is stored energy of terrible events, then the terrible event energy spontaneously reads out from the crystal structure of the fabric. In modern parlance, you could say that the stored energy is suddenly triggered to download into the atmosphere. When humans witness these things, they see them as a surprising and very often frightful event or action, and they record it in their minds as a ghost. In a way, you could say that the energy that is stored and appears as a ghost is so active that it "reads" itself out of the material where it's stored – it downloads itself from the fabric of the building into the atmosphere and mental processes of the human that becomes aware of something strange happening.

That is why in your case of the Lincoln Inn there was a ghost in your room. (Which you remember as Number 21 - better check that in our book of psychic adventures, strange how that book has come up at this particular time when you are talking about this sort of thing, isn't it? – Smile from us!) In that case of the Lincoln Inn, the person that gave rise to the intense vibration stored in the fabric of the walls – stone walls, by the way, so that there was a lot of crystalline material available to record the events – that person was jilted in love. She was living in the building which is now the inn, it was an inn in those days too. From the clothes of the ghost you saw she lived in Victorian times (1837-1901). When you walked into the room, the emanations of energy from the walls molded themselves into the form the woman took as she lay on her deathbed. She was so unhappy, so completely awash in anguish, that she took poison to kill herself, then she lay on the bed to die. As she died, her energy was stored in the fabric of the walls of that particular room. In later times, when anybody went in that room, the stored energy automatically downloaded a reading of itself.

It's like having material in a computer memory, and every time you want to see that material you can print it out and take away a copy, but it doesn't change the stored information at all. It remains in the computer and is not decreased in amount or intensity, and at any time you are able to read it out as a printed sheet of paper. In all these cases where spontaneous appearances occur it's like the computer deciding it's going to print out a page or two of text because it's so important to the originator of the program. It seems that the originator leaves instructions that every time a human pays attention to the program, it should download another print and make an appearance in apparently physical form. The woman you saw lying on the bed as you came in was not made of physical flesh and bones, but was like a hologram of energy, and that hologram had been stored in the walls of the building.

Basically, that is the mechanism of producing a ghost. We can tell you that all the time there is no human around, everything is peaceful, but as soon as the stored energy detects the presence of a human that can receive the message – which is to hear screams or chains rattling or see a vision of a person maybe with a physical deformity relating to the way they died – as soon as the human is detected by the energy, then it automatically downloads the energy performance that humans see as a ghost or hear as ghostly noises.

Summarizing, a ghost is energy that is recorded in the surroundings. That is quite distinct and different to the appearance of a spirit, which is an energy formation that was once in a human body – or an animal body for that matter – it exists in its own right. Even in heaven, before a person is born there is an energy construct that we recognize as a spirit. We are all spirit beings. You humans sometimes take on the physical form of a human to experience and learn. When that experience – your lesson – is over, you return back to the place you call heaven, the spirit world or the 5th dimension. You are not recorded in any situation like a ghost, but you could say your spirit is recorded in the body that you build.

When a spirit is released from the body at death or in an out-of-body experience, it can travel as an energy construct. In certain circumstances, those energy constructs that you call spirits can make appearances to people who are sensitive and they can communicate with a spirit that is in the body, and that of course is what you call a human medium. Those spirits are quite distinct from ghosts that we were talking about earlier. Spirits have to absorb energy from the surroundings to give them energy to move and exist when they are on the Earth plane. Just as you have eaten your breakfast today, Malcolm, to feed your body, you feed your spirit with love. All energy is love energy, really. When a spirit is trapped in a place, for example, your mother still in her house waiting for you to return from Canada, then to exist, to have energy acting as nutrition, that spirit pulls energy from the surroundings in the building where it is staying. When a human comes into that building it detects the energy loss as a cold atmosphere. There is an actual decrease in the temperature of the building, it's not just a physical effect on the body of the perceiving human.

If you look in the book called *The Haunting of Borely Rectory* (Ref.3-1), you will see there was a measurable decrease in temperature. In that book, a ghost hunting group measured the temperature differences at several degrees Fahrenheit as a measure of the amount of energy that was absorbed by the spirit trapped in that location.

The most intense reduction of temperature resulting from spirit absorption of energy was the one you experienced in the haunted inn in the English Lake District. You know which one we mean, Malcolm. There, it was, as you said, like standing in a walk-in refrigerator when you stood in the alcove that had once been the doorway. That was because the spirit that was still trapped there, since coaching times, had needed energy to maintain itself so it subtracted heat energy from the building resulting in very low temperatures in one particular room.

Summarizing the main differences between spirits and ghosts: The energy that is perceived as a ghost is stored in crystalline material in the fabric of the building or in the ground under the event and that energy is read out automatically. It's like printing out a page from a computer memory, since the amount of energy that is stored is not decreased by each reading out. When the energy detects the attention of a nearby human it downloads into the atmosphere and the human perceives the energy as a ghostly form or noise.

In the case of a spirit that is trapped on the Earth plane, maybe as a result of its own need to settle something that happened during its own lifetime, it needs energy to maintain itself in the Earth's atmosphere. So, it removes heat energy from its surroundings, and this is perceived by humans as a deathly chill in the building. If you encounter such spirits, then it's your duty as a human, being of the spirit world, to try to help the trapped spirit be released. You helped the spirit in the Lake District inn by praying for its release and telling it that it could go on to live in the next dimension, as all spirits do after death. In that case, you were partly successful, Malcolm, but other people had to add their assistance in later years since you were there, so it didn't happen right away.

That's all we wanted to tell you, Malcolm. Don't worry about your health, it's improving greatly and it's meant to improve greatly now you have had the experience of the stomach problems, the Parkinson's and the pinched nerve in your back. Those were all events that were meant to test your strength and you have come through all those with flying colours. So, now your health is returning to normal and you will be able to get on with your work.

As for the money to pay for the publication of the new book, don't give any concern to that, it will appear in due course. You will have it ready to pay for the publication just when you need it. All is taken care of, you don't need to do any special effort or work to earn that money, it will be presented to you when the time comes for the publication.

Now we leave you, Malcolm. We feel that this was a very useful additional communication. We didn't expect to have another one with you, but we offered to answer any questions that came up, and Maxine's question to you yesterday about the difference between spirits and ghosts brought back a remembrance to you. It acted as a reminder to you that this was something that you needed to include in our second book, and so we were very happy that you were able to come here and talk for us today channeling this material on ghosts and spirits. Did you see an image in your eyes just then – a head? That was the head of one of the Pleiadians who is around you helping you in your work. Often when you are sitting alone, you hear noises that may be from your Guides – they use taps on the television set and other parts of the room – often when you have a thought puzzle in your head, then you hear other noises in the structure of the building, that is a presence from the Pleiades. They can be with you in less than any of your time units – at the speed of thought.

Now we leave you, Malcolm. Thank you doing this for us. Thank you for your curiosity and your helping of the other people in your world. You will be able to help a lot more people when this book is published, because not only will they read this book and understand some things that have been puzzling them but also it will help publicize our first book. The two books together will become very popular.

We leave you now with our thanks and all the blessings that are ours to give. We assure you that you will have time for a vacation this year. We will help you put our book together and get it ready for publication quite soon.

Our love to you, Malcolm, and our thanks. A happy summer in this beautiful green place you call Vancouver. Goodbye from us all. We love you and we know you love us. Thank you Malcolm and goodbye.

Malcolm K. Smith / June 15, 2012

Appendix 16:
Paper Wicking Experimental Details and Results

"Love" and "Hate" were the words I chose to put on labels which I put on two identical new glass jars with capacity about 1 liter each. I filled the jars with cold tap water "layering" in the water, i.e. first a quarter of a liter was run – straight from the tap – into one jar, the next quarter of a liter into the other jar, then the next quarter of a liter was put in the first jar and so on until they were full. This layering procedure ensured mixing of the water in the two jars and minimized the chances of a systematic difference, such as temperature, between the jars. After being filled, the jars were sealed with an airtight closure device fitted to them.

Next the jar labeled "Love" was put in my car trunk overnight where it "listened" to classical baroque music for 12 hours. The next night the jar labeled "Hate" listened to discordant rock music for 12 hours in the same car trunk. (During the preparation of the samples the car was in a closed garage on consecutive mild nights.)

For the wicking experiments, coffee filter paper was cut into 2 X 10 centimetre strips, each strip was labeled with a number and then it was cut lengthwise into two parallel strips measuring 1 X 10 centimetres. This was done to make the structure of the paper in the strips as similar as possible. Then the strips were marked with pencil lines at 1.5 and 9.5 centimetres from the bottom end of the strip, forming an 8 centimetre "run zone" over which the advancing front of the water could be timed. 1.0 centimetre from the bottom of the strip, a line was drawn across the strip with a pen containing water-soluble colour ink. Finally, another pencil line was drawn across the strip 0.5 centimetre from the bottom – this was the mark to which the paper strip was submerged in the water sample under test.

Two new graduated measuring cylinders marked Love or Hate were used for the wicking experiments. The same volume of treated water was placed in each of the cylinders – always keeping the Love water sample in the Love marked cylinder, etc. The pairs of paper strips were suspended from glass rods placed across the mouth of the cylinders, the strips were clamped by sprung paper clips at a position that dipped the bottom 0.5 centimetre of the paper strip in the water under test.

The water started wicking up the strips, dye from the ink line was picked up by the advancing water front, and as it crossed the first pencil line a stop watch was started and was not stopped until the dye marker crossed the second pencil line 8 centimetres above. Each run took around 7-8 minutes, the time was recorded for each run in a notebook.

Each wicking experiment involved two 1 X 10 strips cut from a 2 X 10 centimetre strip of the same filter paper. The Love and Hate samples were always run in parallel and the times were recorded on separate stop watches. The position of the cylinders relative to the surroundings was reversed after each run to minimize atmospheric effects.

By the method described above, 20 pairs of strips were run with the results shown in Table A1. The results were also plotted in a frequency diagram, Figure A1.

Table A1, Water Wicking Experiments – 2006

Time for water to climb 8 cm strip of filter paper
Love and Hate water samples prepared 27 Aug. – 2 Sept. 2006.

STRIP CODE	HATE WATER SAMPLE	LOVE WATER SAMPLE
Tests 13 Sept. 2006		
1A	467 seconds	534 seconds
1 B	439	464
Tests 18 Sept. 2006		
6B	566	538
6A	466	622
5A	475	506
4A	433	511
19 Sept. 2006		
4B	445	482
3B	408	478
2B	494	463
Total Times	4193	4598
Mean Time	466	511
Difference	9.7%	

Table A1 Continued:

Love and Hate water samples prepared 2 &3 Oct. 2006

STRIP CODE	HATE WATER SAMPLE	LOVE WATER SAMPLE
Tests 4 Oct. 2006		
6A	484 seconds	496 seconds
5B	429	480
5A	429	465
4B	455	470
1B	475	484
Tests 8 Oct. 2006		
6B	561	541
1A	455	475
3B	489	525
2B	521	456
Tests 10 Oct. 2006		
4A	339	471
2A	425	505
Total Times	5062	5368
Mean Time	460	488
Difference	6.1%	

I measured the wicking time for 20 pairs of Love and Hate water. After 9 pairs had been run new batches of Love and Hate water samples were treated by the same music as before and 11 pairs of wicking tests were run on the new water samples.

Summarizing the results:

- The Love water took longer than the Hate water to wick up 8 centimetres of filter paper in 16 out of 20 runs.

- For the first batch of music treated water the average difference in time to wick up 8 centimetres was 9.7% and for the second batch the difference was 6.1%.

The results show clearly that there is a consistent physical difference between the Love- and Hate-treated water samples.

These results are opposite to those expected. As described earlier, in flow through cells, the smaller oligomers, created by love treatment, flow faster through the tiny cell orifices. But in the wicking experiments the love-treated water was slower to flow through the gaps between fibres. However the fibre gaps are thousands of times bigger than the cell orifices. This suggests different mechanisms are operating in the paper fibre gaps and the cell orifices. I looked up the formula (*Van Nostrand's Scientific Encyclopedia*) for capillary rise:

$$h = \frac{2 T \cos \alpha}{r \rho g}$$

(Where h is the height of the liquid capillary rise; r is the internal radius of a circular capillary; α is the contact angle; T is the surface tension; ρ is the liquid density and g is gravity.)

This showed the water viscosity did not influence the pressure driving the wicking. But the pressure was proportional to two times the surface tension and it seems likely that the surface tension will be bigger if the water consists of big oligomers.

So, I conclude from this brief experiment:

- There is a consistent physical difference between water treated with hard rock and gentle classical music.

- The surface tension of the water sample treated with hard rock music appears to be slightly higher than that of the classical music treated water on account of the larger oligomers in the former and this resulted in faster wicking up the strips of filter paper.

Index

A

acquired knowledge 50, 191
agate 8, 21
akashic 173
alien space ship 28, 112
All energy is love 42, 232
All That Is 32, 63, 89, 110, 138, 186, 194, 196, 203
Angel 2, 7, 11–15, 17, 18, 22–25, 28–31, 33–34, 38, 41, 48, 54, 56, 58–62, 66, 72, 77–80, 86, 89, 91, 93, 99, 101, 112–113, 127, 131–134, 136, 138–139, 141, 145, 153, 156–159, 163–169, 171, 173, 176–178, 181, 187, 189, 195, 197–200, 202, 205, 211, 215, 227
apatite 23–24, 31, 111, 128, 130, 136–137, 158, 183–184, 186
apparent viscosity 68, 149, 206
aspect 13, 25, 75, 109, 116, 118, 147–148, 153, 177, 211–212, 215
Atlantean 16–17, 103–104, 174
Atlantis 17, 103, 174
atom 5–6, 8–9, 11–12, 15, 21–24, 34, 46, 48, 58, 65–66, 71, 77–80, 100, 102, 104, 113–114, 137, 145, 160, 190, 202, 217–219, 227

B

battle 35–36, 228–229
bone 6, 16, 21, 23–32, 39, 93–94, 99, 110–118, 128–137, 158, 169, 181, 183–186, 231
brain memory and cellular memory 54, 195

C

capillary rise 70, 239
Carole Wilson – a Canadian psychic and police psychometry expert 18
cartilage 23
cell 5, 10–12, 14–16, 22–24, 31–32, 34, 45–52, 54, 57, 65, 68, 70, 73, 76, 82, 93, 100, 102–105, 107, 109, 112, 114, 127–128, 133–137, 149, 150, 163, 171, 173–175, 177, 184–186, 189–195, 206, 209, 212, 220, 224, 227, 239
cellular memory 51–52, 54, 192, 195
cell wall membrane 47
chains rattling 39, 231
Chi 76, 212
cold atmosphere 41–42, 232
coloured ribbons 55–56, 61, 166–167, 177, 197, 200
connecting with the cosmic lattice 55, 58, 157, 160
cosmic lattice 14, 17, 55–62, 96, 101–103, 105–106, 131–134, 156–160, 163, 165–169, 171–178, 197–198, 199–201
crop formation 65, 72–74, 139, 147–149, 208–210, 213
crystal 2, 5–18, 21, 33–35, 38, 46, 48–49, 52, 56, 66, 75, 79, 89, 91, 97, 100–106, 108–109, 118–119, 127–128, 131–132, 134, 136–138, 148, 150–151, 154, 160–161, 163, 169, 174–175, 177–179, 189–191, 194, 202, 211, 227–229

241

crystal junction 16
crystal lattice 211
Crystal Light 2, 7, 11, 13, 17–18,
    33–34, 38, 48, 56, 66, 89,
    91, 97, 101, 119, 136, 138,
    161, 179, 202
crystalline v, 6–9, 14, 17–18,
    21–24, 35–36, 38, 43, 45,
    47, 50, 55, 67, 75, 107, 111,
    154, 163
crystalline grid 17–18, 171,
    173–176, 178
crystalline sheath 171, 173–178
crystallographer 10
crystals in bone 21, 23–24, 137
crystals in electronics 5, 15
crystal skull 5, 16–19, 132, 174,
    178
crystal structure 5, 8, 11, 34–35,
    38, 52, 104–105, 108–109,
    114, 118, 191, 227–229
crystal unit cell 10, 48–49, 104,
    109, 189–190, 194
crystal versus glass 5, 9
Culloden 36, 229
cytoplasm 47

D

DNA 14, 17, 19, 24–25, 48–50,
    52, 102, 107–108, 111, 114,
    135, 164, 167–169, 173–
    177, 189–191, 193–194
DNA memory retention system 50,
    191

E

electromagnetic song 77, 80–86,
    218–225
energy 3, 5, 7, 9, 11–14, 17–19,
    21–22, 24–28, 30–36, 38–
    45, 48–49, 55– 58, 61–62,
    65–69, 72–78, 81–82,
    84–86, 93–96, 99–106,
    108–112, 114–118, 127–
    133, 135–137, 140–142,
    147–151, 153–155, 163,
    171–172, 174–175, 178,
    181–187, 189–190, 194,
    200–201, 205, 207–213,
    215–217, 220–225, 227–233
energy formation 33, 41–42, 44,
    231
energy stamp 61–62, 72–76,
    81–82, 84–86, 147–149,
    200–201, 207–213, 215–
    216, 220–225
English Lake District 40, 43, 232

F

feeling atmosphere 33, 34
flower 55–56, 59, 155–156, 165,
    169, 177, 197–198
flower method 55, 59, 198
Flying Nun 25, 93, 116
fractal overtone 85, 224
frame 11–12, 31, 61, 100, 184,-
    186
fringed micelle model 22

G

ghost 33–34, 36, 38–39, 41–44,
    227, 229–232, 234
glass 5, 8, 9, 10, 30, 46, 69, 79,
    108, 208, 235, 236
Golden Angel 5, 12–15, 22, 34,
    59–62, 131–134, 136,
    156–159, 163–169, 173,
    176–178, 197–200, 227

H

healing 7, 19, 65, 71, 73, 75–77,
    81–87, 102, 107–109, 137,
    151, 153, 212, 216, 220–225
healing jewelry 77, 83, 85, 224
healing mechanism 77, 84
hexamer 65–69, 72, 206–208

higher self 5, 12–13, 131, 156
hologram 33, 39, 58, 231
homeopathy 7, 65, 72, 81, 84, 147, 155, 158, 205, 207–208, 214, 220, 222
hydrogen bonding 66, 146
hyper-dimensional love energy 56

I

ice crystal 67, 148, 207
impression in sand 167
information energy stored in bones 21, 24
intent 14–15, 49, 67–68, 71, 73, 83, 96, 102–103, 132, 139, 143, 148, 151, 163, 164, 170, 190, 195, 205, 207–208
inter-atomic force 9
interplanetary analysis 77, 80

K

Kirkstone Pass 33, 40

L

Lamarck 50, 51, 190, 192
lattice 9, 14, 17, 55–62, 75, 96, 101–103, 105–106, 131–134, 156–160, 163, 165–169, 171–178, 197–201, 211
Lemuria 17, 174
Lincoln Inn 33, 36, 38, 230
lipid 45–49, 107, 134, 189–190, 194
lipid double layer 45–49, 194
liquid crystal 6, 45–49, 107, 134–137, 189–190, 194
lollipop 46
love energy 30–32, 42, 56, 65, 68–69, 72–74, 76, 85, 94–95, 104, 183–184, 186, 207, 209–210, 212, 215, 224–225, 232

M

magical powers 6, 7
Mars 64, 80, 176, 218–219
metal 1–2, 5, 8, 14, 21, 30, 34, 45, 106, 153–154, 158–160, 163–164, 168–169, 177–178
mineral 5, 7, 8, 23–24, 34, 36, 45, 77, 81, 83–86, 137, 215–217, 219–224, 228–229
molecular formation 73, 208
molecule 5, 6, 9–11, 21–23, 46, 47–48, 58, 65–69, 72–85, 108, 134, 145–150, 173, 175, 189–190, 194, 205–213, 216,–225
morphic field 58, 106–107, 156, 160, 173
multidimensional 14, 17, 167–169, 174
Musical Notation 5
Musical Water Experiment 65, 69
music notation 11–12, 127

N

Neville 40–41
new generation of cells 51, 192–193
Nordic aliens 29, 113
nucleate 8

O

obsidian 8–9
oligomer 66–68, 70–72, 205–207, 216, 239–240
organ transplant 45, 51, 191
Orion 31, 129, 184–186
oversoul 156–157
Oversoul 5, 12–13, 19

P

pendant 1, 2, 14, 153–155, 158–160, 163–164, 168–169, 177–178, 185

243

phospholipid 46–47
piezoelectric 15
pilgrim 27–28, 30, 32, 95, 115–116, 118, 130, 181, 186
pilgrimage 21, 27–28, 95, 115–116, 118, 128, 130, 184
plant 22, 55, 58–62, 67–69, 71–73, 130, 133, 149, 156–157, 160, 165–169, 173, 177–178, 189, 198–201, 206–207, 209
Pleiades 17, 174, 234
polymer 22
Prana 76, 212
psychometric 14–15, 22, 55–56, 59, 60–62, 154, 158–159, 163–164, 166–169, 199
psychometric technique 55–56, 59, 61–62, 159
psychometrist 165–166
psychometry 1, 2, 12–15, 18, 34, 55, 59, 61, 132–133, 153–159, 163–169, 177–178, 185, 197–200, 205, 227

Q

Qi 76
Qi Gong 76, 212, 225
quartz 6, 7, 9, 15–16, 21, 178

R

readee 56, 61, 62, 200–201
reader 15, 48, 56, 59, 61–62, 86, 164–169, 177– 178, 190, 194, 197–198, 200
reading the energy 5, 13–14, 154, 163

S

Saint Teresa of Avila 25, 116
salt 9, 10, 75, 81, 84, 215–216, 220, 222

sand impression 55, 61–62, 168, 177, 197, 200
Santiago de Compostela 21, 27–31, 95–96, 112–113, 115, 129–130, 136, 181, 183–184, 186
scalar energy 85, 106, 224–225
screams 39, 231
semi-crystalline 6, 21–22, 24, 45
sodium chloride 10, 75–76, 82, 105, 211–212, 216, 221
spectra 77–78, 80, 146, 224
spirit 14, 24–26, 32–34, 40–44, 57, 93–94, 111, 114, 116, 132, 158, 164, 176, 181–183, 186–187, 203, 227, 231–234
spirit energy 33, 40–41
Spirit Guide 2
spiritual energy 3, 5, 18, 19
storage of spiritual energy 5
stored information accompanies organ transplant 45, 51
storing energy 5, 11, 21, 24, 45, 49, 108, 194
structure of physical materials 5
surface tension 70, 71, 158, 239, 240
synchronicity 31, 57, 185, 186

T

teeth 23–24, 118, 136
telepathic communication 59, 198
telepathy 57, 172, 176
tetragonal 11
torture 35, 228
tourmaline 7, 83, 149, 150, 216–217, 219, 221
transfer of acquired information in the animal world 45, 53
transistor 15–16
triclinic 11
tsunami 172
types of human bone tissue 23

U

UFO 28, 30, 112
unit cell 5, 10–12, 14–16, 22, 24, 31, 34, 48, 49, 65, 100, 103–105, 109, 114, 127–128, 133–135, 137, 163, 184–186, 189–190, 194, 227

W

water as a recording medium 65, 71, 73, 216
Water Chemistry 65
whisker crystal 79–80, 218
wicking time 70, 126, 239
wood 6, 22–23, 35, 45, 198, 228

'Science doesn't make sense without God' P. 63.